Mute Vol 2 #12

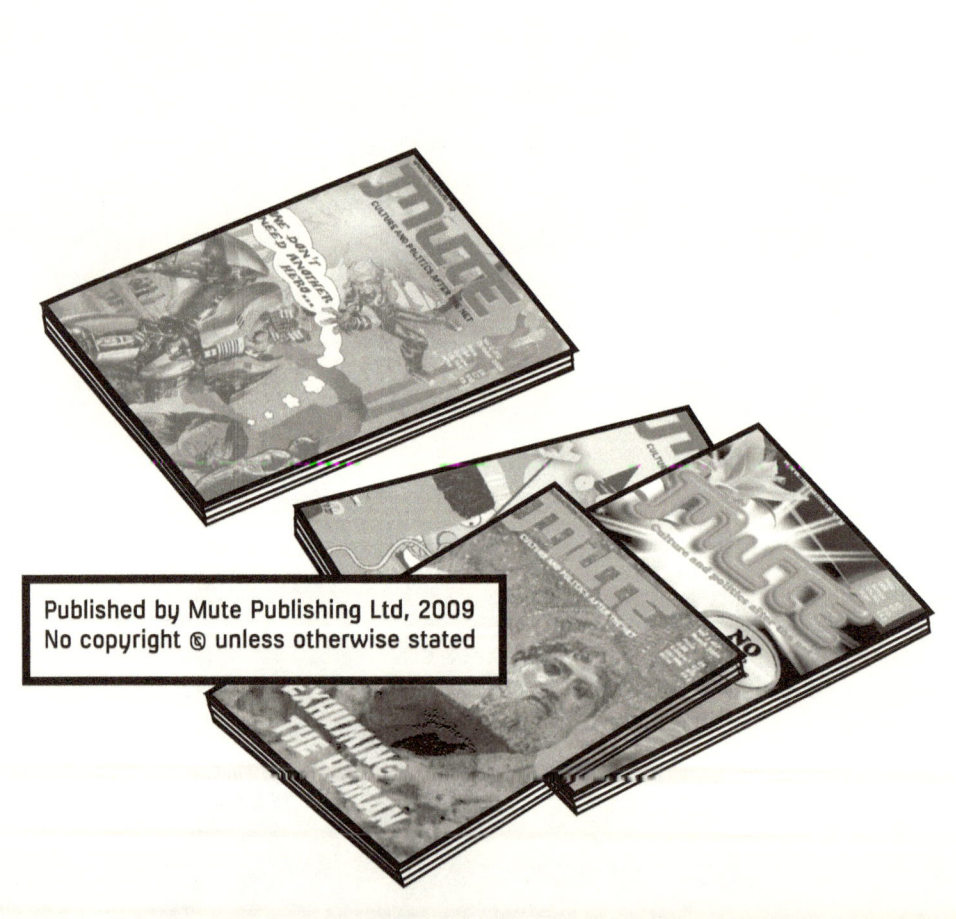

Published by Mute Publishing Ltd, 2009
No copyright ⓒ unless otherwise stated

MUTE Vol 2 #12
SUMMER ISSUE – JUNE '09

EDITOR
Josephine Berry Slater <josie@metamute.org>

EDITORIAL BOARD
Josephine Berry Slater, Matthew Hyland <infuriant@autistici.org>, Anthony Iles <anthony@metamute.org>, Demetra Kotouza <demetra@inventati.org>, Hari Kunzru <hari@metamute.org>, Melancholic Troglodytes <meltrogs1@hotmail.com>, Pauline van Mourik Broekman, Benedict Seymour <ben@metamute.org>, Stefan Szczelkun <szczels@ukonline.co.uk> and Simon Worthington

MUTE PUBLISHING ADVISORY BOARD
Ceri Hand, Sally Jane Norman, Sukhdev Sandhu and Andy Wilson

PUBLISHERS
Pauline van Mourik Broekman <pauline@metamute.org>
Simon Worthington <simon@metamute.org>

ISSUE DESIGN
Laura Oldenbourg <laura@metamute.org>

ADVERTISING & MARKETING
Lois Olmstead <lois@metamute.org>
T: +44 (0)7791284039

WEBSITE
Metamute.org is powered by Drupal and CiviCRM FLOSS Software, with additional software services by our very own OpenMute http://openmute.org

TECH SUPPORT
Web infrastructure: Darron Broad <darron@kewl.org>

PROJECT ASSISTANT CO-ORDINATOR
Caroline Heron <caroline@metamute.org>

INTERNS
Paul Graham, Fahima Haque and Olga Panadés

OFFICE
Mute, Unit 9, The Whitechapel Centre,
85 Myrdle Street,
London E1 1HQ, UK
T: +44 (0)20 7377 6949
F: +44 (0)20 7377 9520
email: <mute@metamute.org>

SUBSCRIPTIONS
Howard Slater
T: +44 (0)20 7377 6949
F: +44 (0)20 7377 9520
email: <subs@metamute.org>
web: http://www.metamute.org/subs/

DISTRIBUTION UK
Central Books,
99 Wallis Road,
London, E4 5LN
T: +44 (0)20 8986 4854
F: +44 (0)20 8533 5821

DISTRIBUTION NORTH AMERICA
Please contact:
Lois Olmstead <lois@metamute.org>
T: +44 (0)7791284039
or visit http://www.moreismore.net

CONTRIBUTING
Mute welcomes contributions of all kinds. Email <mute@metamute.org> with your ideas

You can also publish on Mute's website [http://www.metamute.org]. Post news, texts, events and comments, or upload media to the Mute Public Library http://pl.metamute.org

The views expressed in Mute and Metamute are not necessarily those of the publishers or service providers

Mute is published in the UK by Mute Publishing Ltd. and printed by OpenMute [http://openmute.org] print on demand (POD) book services in the USA and UK

COVER
Nils Norman <nils@dismalgarden.com>

SPECIAL THANKS
To Nils Norman for stripping the flesh from the bones of the creative class with such speed and relish. And to Howard for extra proofing

ISSN 1356-7748 - 212
ISBN 978-1-906496-34-0

Mute is supported by
Arts Council England

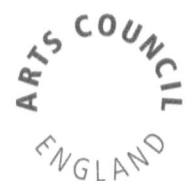

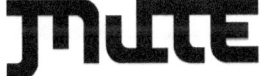

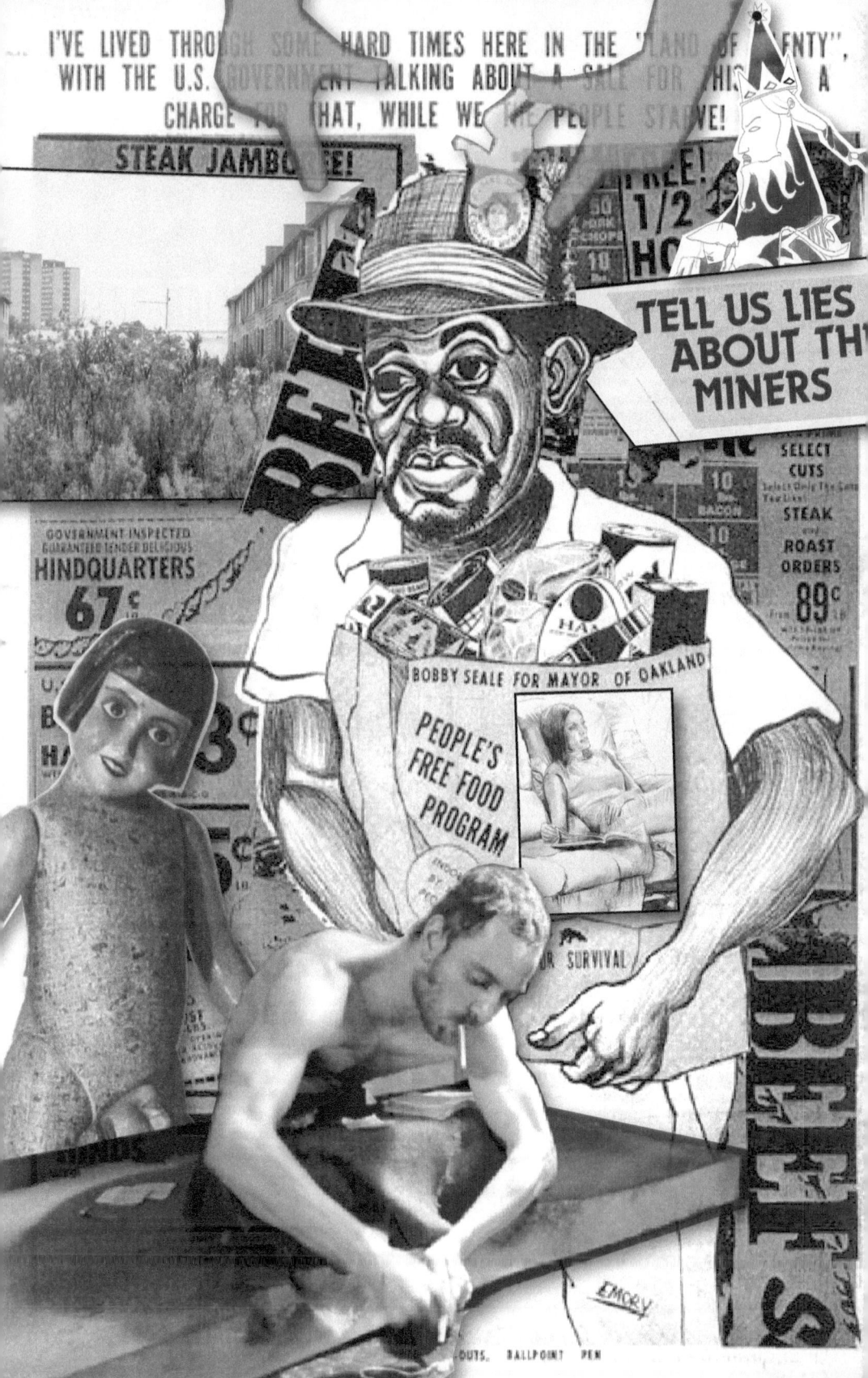

CONTENTS

6   **EDITORIAL**
by Josephine Berry Slater

14   **DÉRIVING UNDER THE INFLUENCE**
Chris Jones inspects the wounds opened by Laura Oldfield Ford's pictures of regenerate London

24   **CG2014: FORMULARY FOR A SKEWED URBANISM**
Neil Gray ambushes the cowboy capitalists staking out Glasgow's 'urban frontier'

38   **THE CREATIVE CITY IN RUINS**
Artist's project by Nils Norman

44   **CONCERNING ART AND SOCIAL CHANGE**
Brian Holmes and Marco Deseriis on critical culture within recuperative 'semiocapitalism'

58   **ALL MOUTH, NO HISTORY**
William Dixon gets gobby with Christian Marazzi and his linguistic analysis of financialisation

64   **DEBT: THE FIRST FIVE THOUSAND YEARS**
David Graeber gives us the elevator pitch on debt's violent history

76   **HUNGRY GHOST**
Steve McQueen's film *Hunger* whets Paul Helliwell's appetite for some political context

86   **A CLIMATIC DISORDER?**
John Cunningham clears the air after a meeting between Climate Campers and the NUM

94   **THE SIMPLE EXPRESSION OF COMPLEX THOUGHT**
M. Beatrice Fazi splices interactive media and the philosophy of expression

106   **OBJECTIVE PHANTOMS**
Kenneth Cox toys with Romanian poet Ghérasim Luca's objects and desires

EDITORIAL

'The world', writes Deleuze, 'does not exist outside its expressions'. This pivotal quote in M. Beatrice Fazi's piece, 'The Simple Expression of Complex Thought'(p.94), helps her to assert a 'realist metaphysics' of expression critical of postmodernism and its readings of the expressive subject. In a rather large nutshell, Fazi is arguing that 'constructivist' readings of phenomena – socio-technical systems, aesthetics, culture – are still mired in a logic of causality. While acknowledging that the subject is contingent and produced, postmodernism nevertheless sees expression as the registering of a prior, inner experience. But expression, argues Fazi in line with Deleuze, is not the transposition of an inner life to an external symbolic system or material, but rather an immanent and differential activity by which the world speaks itself through the interactions of its parts. If we take expression to be the effect of differential sets of relations which do not necessarily entail willed communication, then we should try to see the recent outbreak of the A/H1N1 virus, or swine fever, as a form of expression and not just a mechanistic, viral event.

The contemporary, industrialised world, we could say, does not exist outside its *pathological* expressions – global warming, species extinction, avian and now swine flu. Thinking about such events in expressive terms can be helpful to understanding the deep continuities between abstract and real phenomena, capital and its physical expressions. In this sense, that the potential pandemic follows hard on the heels of the deepest financial crisis the world has seen since the '20s/'30s cannot be shrugged off as coincidence. This idea becomes more striking when we consider, as a homeopath friend once pointed out, the highly related historical sequence of WWI with its huge mobilisation, compression and suffering of bodies, followed by the Spanish flu pandemic of 1918 and then the Wall St. Crash of 1929. As Fazi might agree, this is no Gaia-like expression of the world's internal suffering, but the proliferating iteration of myriad systems of social production and destruction in the age of industrial capitalism.

On the subject of DNA, Fazi says, 'it is abstract yet real, having concrete effects and emerging physicality.' It is not that a code imposes itself onto an inert, available form, but that it is capable of generating form – it is 'morphogenetically pregnant'. Similarly, the immanent activity of abstract capital – its infinite need to accumulate, drive down productive costs, intensify the forces of production and expel labour from the system, etc. – has an emergent physical form. The drive to provide competitively priced meat on a mass scale to satisfy (and stimulate) the new dietary needs of the world's urban majority demands 'the scientific management of poultry and livestock based on principles of industrialized production', Will Barnes has written recently on the Meltdown mailinglist. This, he continues, is 'generating a qualitative amplification of the incidence and scope of disease by massing literally thousands of millions of birds or mammals, respectively, in tightly enclosed quarters ... where naturally occurring disease can run, uninhibited, through these populations at frighteningly rapid speeds creating virally mutating highly pathogenic, potentially pandemic viruses.'

Barnes also states that Mexico City, the place where the swine flu first broke out on a large scale, constitutes the 'classical locus' of a pandemic. The conditions of overcrowding and lack of hygiene in the city's slums are equivalent, in virological terms, to contemporary, industrialised animal farms, slaughter houses and the trenches of WWI. Expressive as the virus is of the widespread capture of human beings and animals into squalid and life threatening conditions under the percussive waves of late capitalist land clearance and enclosures, it is ironic to hear some describe the outbreak as a state-media orchestrated *distraction* from the financial crisis. While pundits refer to a real economy underlying the financial sector, there is no equivalent acknowledgement of real bodies and life forms registering the affects of out-of-control financialisation and over accumulation. Given the refusal of 'consensus reality' to take seriously the *expressive* content of the pandemics and ecological disasters that are breaking out with breathtaking regularity, it is a further irony that expressivity and creativity are fetishised as never before by post-Fordist state planners and the entrepreneuriat.

As well we know, this creative expressivity is so narrowly defined as to be utterly meaningless. In this regime of instrumentalised aesthetics, the inflated sensibilities of the creative few are draped, fig-leaf like, over the narrowing creative possibilities of the dehumanised majority. In his article on the regeneration of Glasgow (p.24), Neil Gray reveals the depth of planners', policy drafters' and their media allies' scorn for that majority, with East End inhabitants described with undisguised class hatred: 'The people do not look good here. Often it is difficult to tell men from women, old men from older men [...] the locals have the blotchy pallor of cave-dwelling consumptives.' Any notion that the creative city might entail some trace of redistributive aspiration should be put to rest by statements such as these – as if we really needed such evidence to see through this urban strategic sham of public subsidy and private looting. Nils Normans' picture story (p.38) conflates the vampirism of the creative city model with that of a decadent art market, as creativity and living labour are sucked dry by the sickly parasite of capital. The jumble of tents running across these images are both apocalyptic – the overcrowding of slums and meat factories erupting in the heart of the Western metropolis – and amusing. These tents are the slick, new purchases of home-counties festival goers and Halfords shoppers. The suburbanites have been deprived of their luxurious space and are compressed into a camp-like jumble; the evacuees from paradise in new, proximate relations to the Creative City's excluded. A new distribution with new, expressive potentials perhaps?

Josephine Berry Slater <josie@metamute.org> is Editor of *Mute*

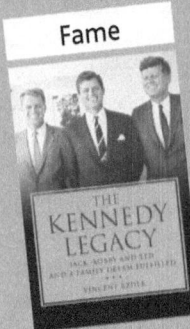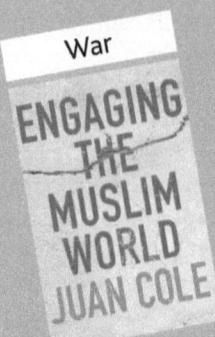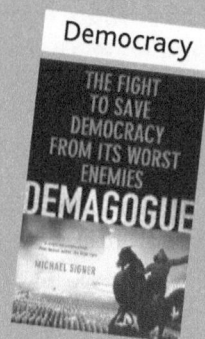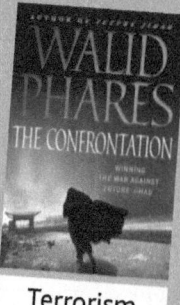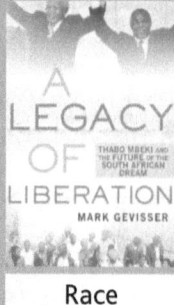

The MIT Press
http://mitpress.mit.edu

Fitzroy House, 11 Chenies Street, London WC1E 7EY
tel: 020 7306 0603 • orders: 01243 779 777

The Monstrosity of Christ
Paradox or Dialectic?
SLAVOJ ŽIŽEK AND JOHN MILBANK
EDITED BY CRESTON DAVIS

A debate between an atheist who recognises the emancipatory potential offered by Christian theology and a theologian who sees global capitalism as the 21st century's greatest ethical challenge. Topics under discussion include truth, subjectivity, transcendence, and the future of religion, secularity and political hope in the light of a monsterful event — God becoming human.

£18.95 • cloth • 320 pp. • 978-0-262-01271-3

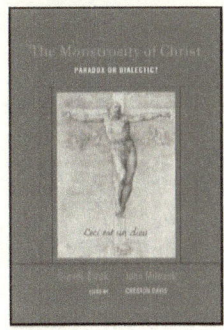

Distributed for Semiotext(e)
Terror from the Air
PETER SLOTERDIJK
TRANSLATED BY AMY PATTON AND STEVE CORCORAN

This book looks at how warfare has changed, attacking not just an enemy's body, but their environment. From the first use of chlorine gas at Ypres, to the toxic gas used at Auschwitz, from the bombing of Dresden to the attack on the World Trade Center, "terror from the air" attacks the very conditions necessary for life.

£9.95 • paper • 128 pp. • 978-1-58435-072-9

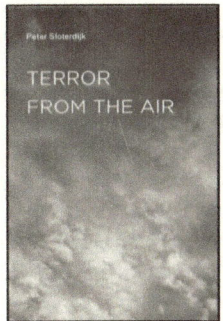

Conservation Refugees
The Hundred-Year Conflict between Global Conservation and Native Peoples
MARK DOWIE

"...[An] important book that illuminates the dark side of the heroic profile of global conservation NGOs: biodiversity conservation has often been achieved, or at least attempted, at the expense of further impoverishing some of the poorest people on the planet...In a sign of sea change, some ecologists are beginning to accept that this brutal exclusion was not always necessary for conservation ends, and that people and nature can survive and flourish together."
— **David Bray,** Department of Environmental Studies, Florida International University

£18.95 • cloth • 371 pp. • 978-0-262-01261-4

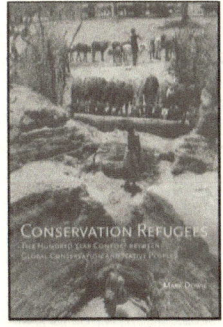

Virtualpolitik
An Electronic History of Government Media-Making in a Time of War, Scandal, Disaster, Miscommunication, and Mistakes
ELIZABETH LOSH

"With incisive scrutiny and careful skepticism, Losh adds a new and crucial arrow to the digital rhetorician's quiver. Governmental uses of digital media reveal official positions on digital and real life alike, positions frequently characterised by blunder, oversight, and delusion. We often hear individuals, businesses, and organisations tout the ways digital media like blogs, videogames, and online video can change culture for the better. Virtualpolitik is a must read for anyone interested in understanding the impediments to such a future."
— **Ian Bogost,** The Georgia Institute of Technology, author of *Persuasive Games*

£19.95 • cloth • 416 pp. (71 illus.) • 978-0-262-12304-4

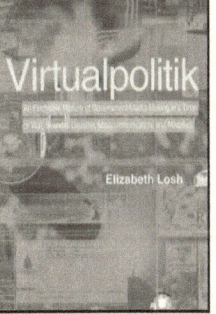

Mute and Autonomedia are pleased to announce the publication of

Proud to be Flesh: A Mute Magazine Anthology of Cultural Politics after the Net

Edited by Josephine Berry Slater and Pauline van Mourik Broekman with Michael Corris, Anthony Iles, Benedict Seymour and Simon Worthington

Nearly 15 years in the making, Mute's forthcoming, supersized anthology offers some of our finest articles thematically organised around key contemporary issues. Featuring seminal writing on digital aesthetics, political art, Web 2.0, the politics of globalisation, free culture and the knowledge commons, *Proud to Be Flesh* takes on 'culture and politics after the net' with relentless intelligence, originality and passion.

Available for pre-order as a limited edition, full-colour hardcover. 624 pages of Mute's best writing, artwork and design with 48 pages of colour illustrations.

Pre-order price: £45.99 + p&p
Publishing date Sept 2009

Pre-order and book preview:
metamute.org/proudtobeflesh

For enquiries contact Lois at lois@metamute.org or call +44 (0)20 7377 6949 Skype mute.london

ISBN 978-1-906496-27-2 hardcover

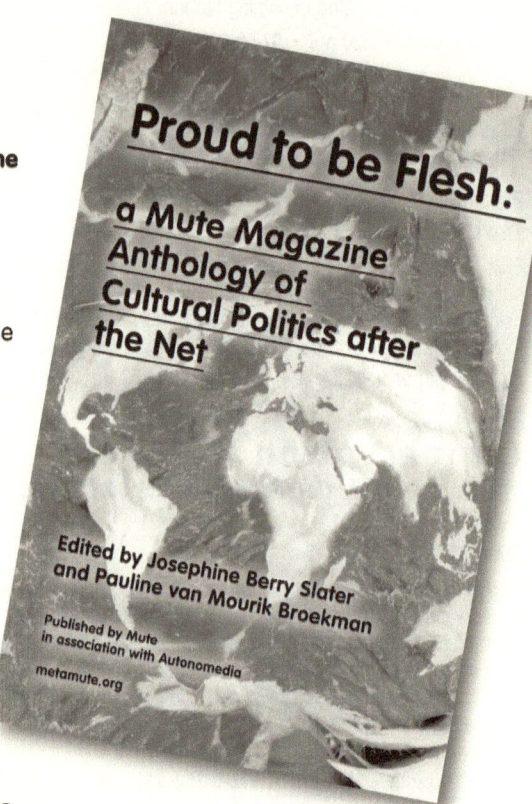

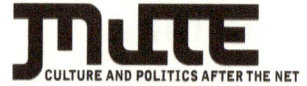

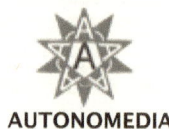

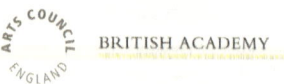

Published by Mute in association with Autonomedia. Supported by Arts Council England and the British Academy

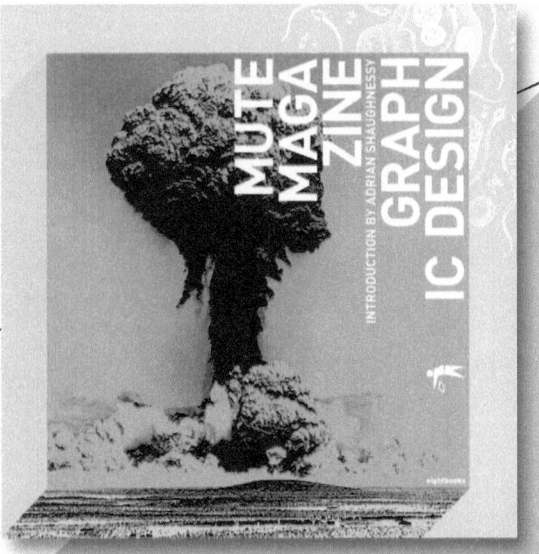
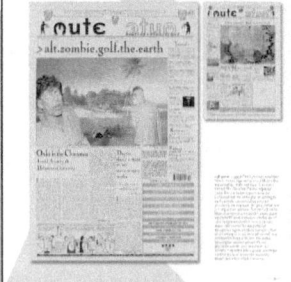
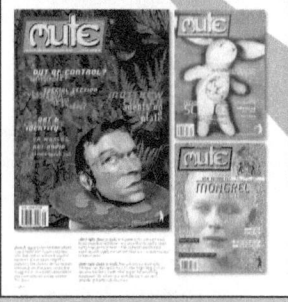

Mute Magazine: Graphic Design

In the early 1990s, long before the internet became an integral part of life, a handful of pioneering magazines took it upon themselves to imagine the web into existence. Using fiction, interviews, speculative theory and experimental graphic design, London-based Mute wielded an influence disproportionate to its scale. Nearly fifteen years after its launch in November 1994, Mute's publication history defines an era, telling the fascinating tale of one publisher's relationship with the 'digital revolution'. This graphic design history presents a full overview of Mute's output, including logos, covers and spreads.

Introduction by Adrian Shaughnessy, with further contributions from Damian Jaques, Pauline van Mourik Broekman and Simon Worthington.

Published by 8books

Softback 220 x 220 mm, 144 pages,
250 colour Illustrations

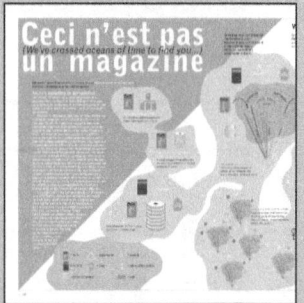
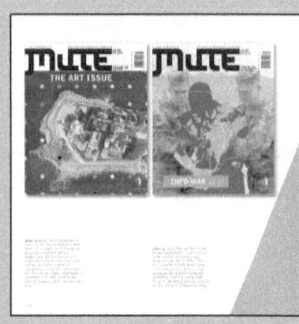

Buy it online at:
metamute.org/mutegraphics

£ 19.95 + p&p

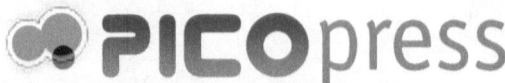
PICOpress

Digital print on demand books

Picopress POD allows you to print books in numbers from 1 to 100 + starting at just £1.99* each. Compared to conventional printing, the set up fees are small too, and instead of financing your whole run in one go, POD allows you to pay as you print.

PicoPress services include:
* POD book printing
* Ecommerce and delivery/ fulfillment services
* POD integration into your existing web site
* Amazon/ Paypal & Google services and integration
* Consultancy and training
* Design and editorial

POD print handling and initial consultation £175 (first publication)

* Consultation on print options and pricing
* Provision of support
* Print handling. This means taking your files and guiding them through the web submissions/revisions process of which the basic steps are:
    o Sending a PDF to the printer
    o Getting the printer to supply you with a proof
    o Publishing the title
    o Creating a library entry on Amazon

picopress.net
info@picopress.net   +44 (0)20 7377 6949

Supported by London Innovations fund
 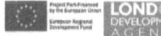

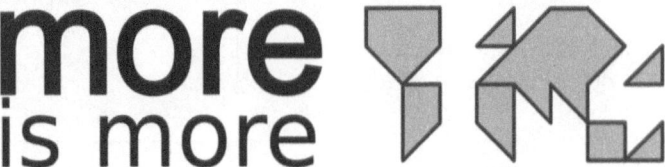

Independent media distribution & video screening network

More is More is an open source, online distribution system for small-scale and independent media. The aim of the network is to provide independent media producers and cultural organisations with a platform that can connect them to local outlets and events. More is More facilitates the sale of goods at such locations as well as direct through the website itself.

Commercial distributors are not best geared to the distribution of media products from the cultural, non-profit or political sectors. OpenMute's distribution network is an attempt to develop an alternative. More is More distributes the following: video, magazines, books, comics, posters, flyers and music. It is also possible to arrange your own event or film-screening through the platform.

While the site is at an alpha stage, we are looking for reactions and input from individual media-producers, cultural and activist organisations as well as a variety of outlets that might be interested in putting their products online or selling them locally.

An OpenMute project

Supported by Digital Pioneers
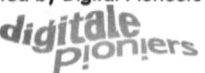

moreismore.net

metamute

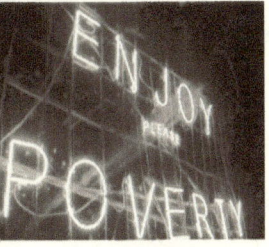

The Atrocity Exhibition
by John Douglas Millar
http://www.metamute.org/en/content/the_atrocity_exhibition
Episode III: 'Enjoy Poverty' is a troubling film about the Congo by Dutch artist Renzo Martens. His 'modest proposal' exposes the inescapable violence that representations of suffering inflict on their subjects. Review by John Douglas Millar

Dissident Island Discs
by Anthony Iles
http://www.metamute.org/en/content/dissident_island_discs
Stefan Szczelkun's online project, 'Agit Disco', invites guest 'DJs' to extract the politics from their music collections into mix CDs. How does this activist archiving add to the inherent politics of popular musics and their reception? - asks Anthony Iles

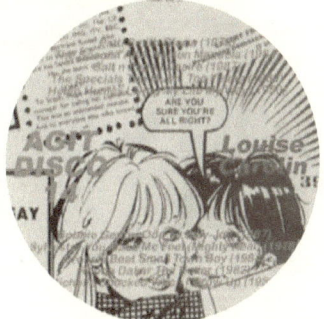

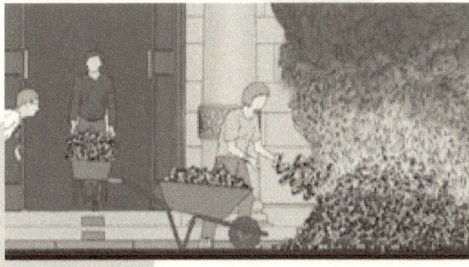

Turning Software Inside Out
by Tony D. Sampson
http://www.metamute.org/en/content/turning_software_inside_out
As our familiarity with software deepens, the question of its cultural understanding looms. Here Tony Sampson reviews *FLOSS+Art and Software Studies: A Lexicon*, two recent books which attempt to open up the black box to a wider audience

From News & Analysis:

Violence in the Name of Respectability: Raids on Activist Spaces During the G20 Summit
by Rampart Collective
http://www.metamute.org/en/violence_in_the_name_of_respectability_raids_on_activist_spaces_during_the_g20_summit
On the Thursday following the G20 protests, two squatted social centres in East London were raided by riot police, apparently looking for instigators of the attacks on the Royal Bank of Scotland on the 1st of April. RampART Social Centre, which has existed for more than four years, and a newly opened Convergence Centre in Earl Street were both being used to house and feed protesters throughout the period of the G20 summit. [...] Other than a brief report in the Independent which referred to unwarranted violence, the raids remained largely unreported.

http://metamute.org

DÉRIVING UNDER THE INFLUENCE

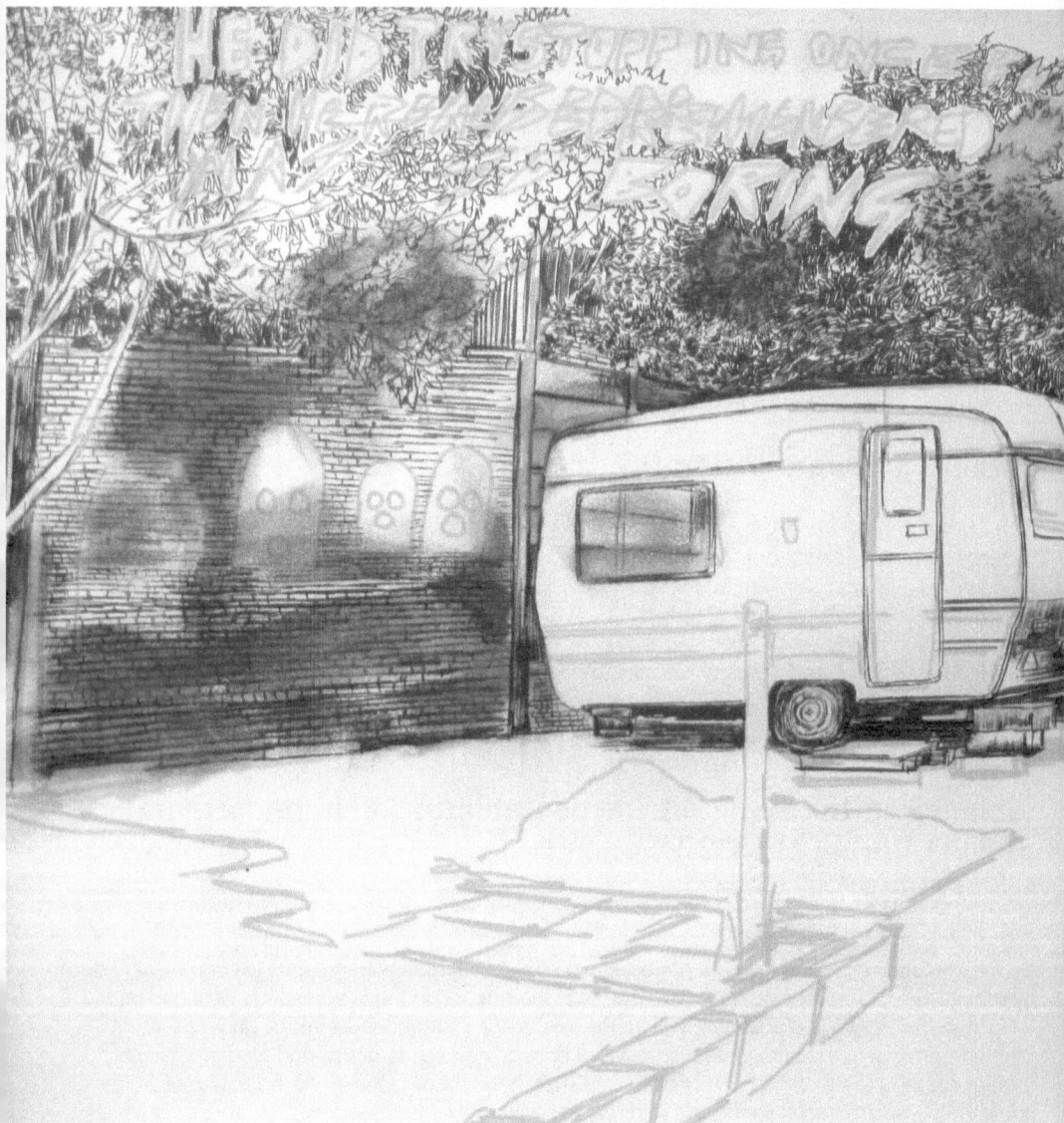

All images: Laura Oldfield Ford, *London 2013, Drifting Through the Ruins*, 2009

Chris Jones

Colliding anarchistic subcultures, zombified yuppies and the ruins of the welfare state, Laura Oldfield Ford's work opens up the economic and cultural wounds of London's regeneration. Review by Chris Jones

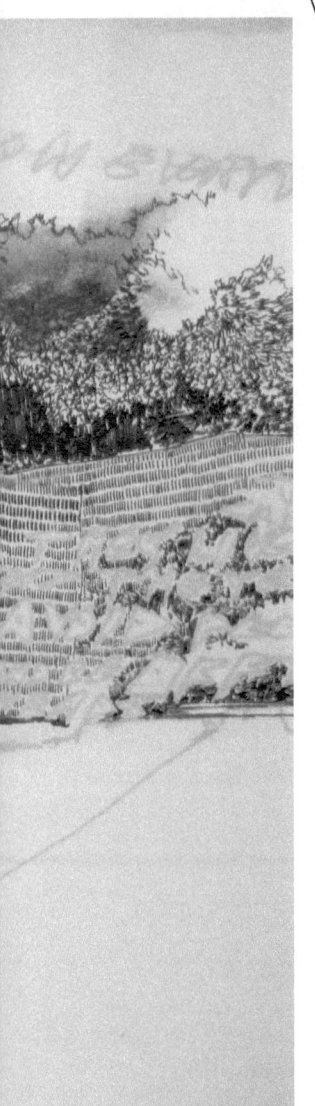

Spitalfields has been described as London's first industrial suburb. But now The City is moving outwards. It wants the land of outcasts for itself.

– Charlie Forman, *Spitalfields: A Battle For Land* (1989)

Brick Lane holds a particular, disorientating memory for me. When stumbling as a teenager through off-my-map streets of London, it was at Brick Lane that I first saw what real poverty was. Not the prosaic suburban poverty of where I was brought up, but the everyday poverty of grim survival. Away from the market proper, at the point where Bethnal Green Road rounds into Shoreditch High St., people had spread stuff out on the pavement and were attempting to make a few quid. People from all nations hawked stuff from all over. Here were heaps of rescued rubbish re-imagined as goods: salvaged nails, tatty hardback books, gone off saucepans.

This feeling of disorientation came back to me when I went to Laura Oldfield Ford's exhibition, London 2013, Drifting Through the Ruins. It doesn't take a book to know that I was walking through an area over which the war had been lost, but it helps. Charlie Forman's *Spitalfields: A Battle for Land* details the fight of local Bangladeshi people against a wave of City-led 'regeneration' for decent housing in the communities where they already lived. Despite some victories, the locals were cast out and the community's businesses moved on or closed. Now the area has been developed with over-glassed high-rises and endless shopping. Spitalfields, once a beautiful place to just walk around in reverie, slowly begins to look like any other mega-complex from New Labour's alleged new economy.

Walking up from London Bridge, I arrive once more where Bethnal Green Road rounds into Shoreditch High St. and ask to be buzzed in to the private space of Hales Gallery.

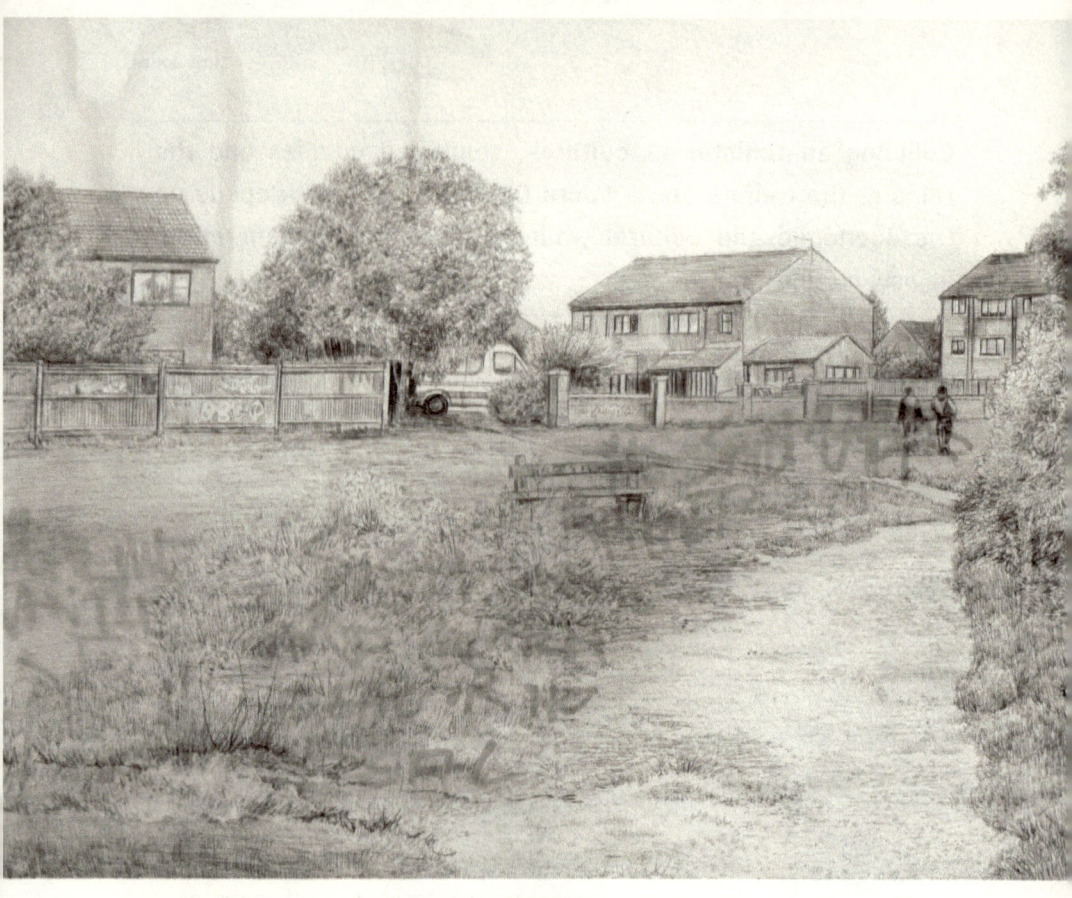

Banged Up: London 2013, Drifting Through the Ruins

Inside the outcasts have come back. The ruins have come back. But they are back as images. One wall of the gallery space has been given up to over 100 of Laura Oldfield Ford's drawings. Here are evocations of time spent wandering in the ruins of the East End after a failed 2012 Olympics. Each image is finely realistic, copied from photos, and so places are familiar to anyone who has been around the London block once or twice. Displayed in tight formation five high by a dozen or so wide, all drawings are the same size; you have to find your own way in. In a conventional fashion, I moved from left to right and looked into each drawing to discern its details and depth.

    These images are not simple landscapes that show us what Oldfield Ford has seen, but intricate montages of places and people that often continue from one sheet to the next. There are views of brutalist towers, but older more poignant local details stand out: old iron and brickwork, dead industrial chimneys, the flow of water through canals, wooden pallets stacked on the wharf's side or the dark fast

Chris Jones

tides of the Thames alongside old tyres on a jetty. Trees and pylons, corrugated iron fencing, metal bollards, dense foliage – all this approaches something like an urban pastoral. These are interzonal territories that fall between the twin evils of work time and leisure. If you climb over fences onto waste ground you will probably find the perfect place to skive, to rethink urbanity and to come down slowly from the rat race you have left behind (albeit temporarily). I like the sad image of a caravan that crops up amidst the ruins, ready and waiting for a cheap holiday in your own misery, the perfect proletarian getaway vehicle.

trees and pylons, corrugated iron fencing, metal bollards, dense foliage – an urban pastoral

It's in these pictured sites of dereliction that I imagine all-night raves or punk squats and secret meetings around bonfires, the passing of the bottle – or more so a spliff – because we had needed to create adult gang huts or play dens among the debris to keep ourselves alive, to ward off our impending, collective zombification. These are wasted spaces only if looked at from the perspective of money making, for these are sites where notions of affluence are contested. Better to have time to explore the ruins than to have money to spend in Ikea.

In the midst of the wastelands and the falling down communities, there is also a population but it's a quiet one. In fact some of the pictures seem like film stills or scenes from short stories. Here two men cross an empty space. Here a woman in a hood stares out of the frame. We don't know what happens next. Not only can we not work out these stories, but nobody is telling either. This collection of pictures of overly-public spaces seems oddly mute and devoid of sound despite them being feral places that seek conversation. Even a bunch of punks in a high street standing by an out-of-place fairground ride, all studs, hairdos and painted leathers seem like ghosts where we would expect noise and movement.

There is a deliberate tension placed between Oldfield Ford's images of ruins and working people, and her depiction of yuppies. The yuppies are almost parodies of themselves, if that's possible. Copied from advert-paeans to luxury and wealth and domestic security, (a woman lounging on a sofa with a glass of wine in hand as if the sofa was the only partner she ever needed, a couple – she with too much hair takes a cocktail from him with too many teeth), these are pictures of people caught in a state of grotesque, self-conscious playfulness brought on by the presence of too much money. But these aspirational types are not left in peace for long; slogans and sentences rise out of the drawings: 'Beware the Mindfuckers', 'Loot Asda, Burn Barratts', 'Wrecker Vandals Criminals St. Modwen' – St. Modwen being a builder of yuppie enclaves.

From Jago to Day-Glo and Back Again

Tower blocks, concrete jungles and UK decay were the signal aesthetic of punk when it crashed the late 1970s culture party. Punk has influenced Laura Oldfield Ford a great deal. Not only does the style continue in her past and present drawings, but you get the sense that the essential political project of autonomy and doing-it-yourself remains just as important. Using the wastelands to hide from the ravages of the economy, punks, squatters, urban nomads, sufferers, ravers, tramps, outcasts all utilise these inner city desert islands to re-invent and experiment with free spaces.

This sense of what you could call 'ruinism' is strongest when you sit back from the wall of images and look upon the collection in its entirety. Although these drawings are made in black biro on white art paper, smudges and stains of Day-Glo pink, yellow and red are streaked across many of the drawings. To me, this summons up D.I.Y. punk fashion's early ludic call for festival among the dreary ruins before punk became only about black and white. But then this is also the colour of Day-Glo signs in the High Street that shriek SALE! SALE! SALE! as everything gets slowly commodified – punk, anarchy, free spaces.

Standing back from it all and viewing it as a whole is rewarding. In its totality, with colours blurring across the swathes of decrepitude, you could convince yourself that another country had suddenly been revealed to you alone. The old and familiar landscapes turn warmly unfamiliar, and zones of escape are suddenly opened up. It would be here that we can see Oldfield Ford's explicit reference and use of the Situationists' early practice of *dérive* and psychogeography. That is, to drift (*dérive*) from place to place, moving not out of the desire to get from A to B but through the emotional influence of ambiance, danger, joy, chance encounter and so on. We cannot drift here and now through these images alone, but some sense of possibility can be gained from the insistence that another life is possible in self-created, autonomous zones, where dereliction can be viewed as a safe haven from the ever-expanding banality of the yuppie territories.

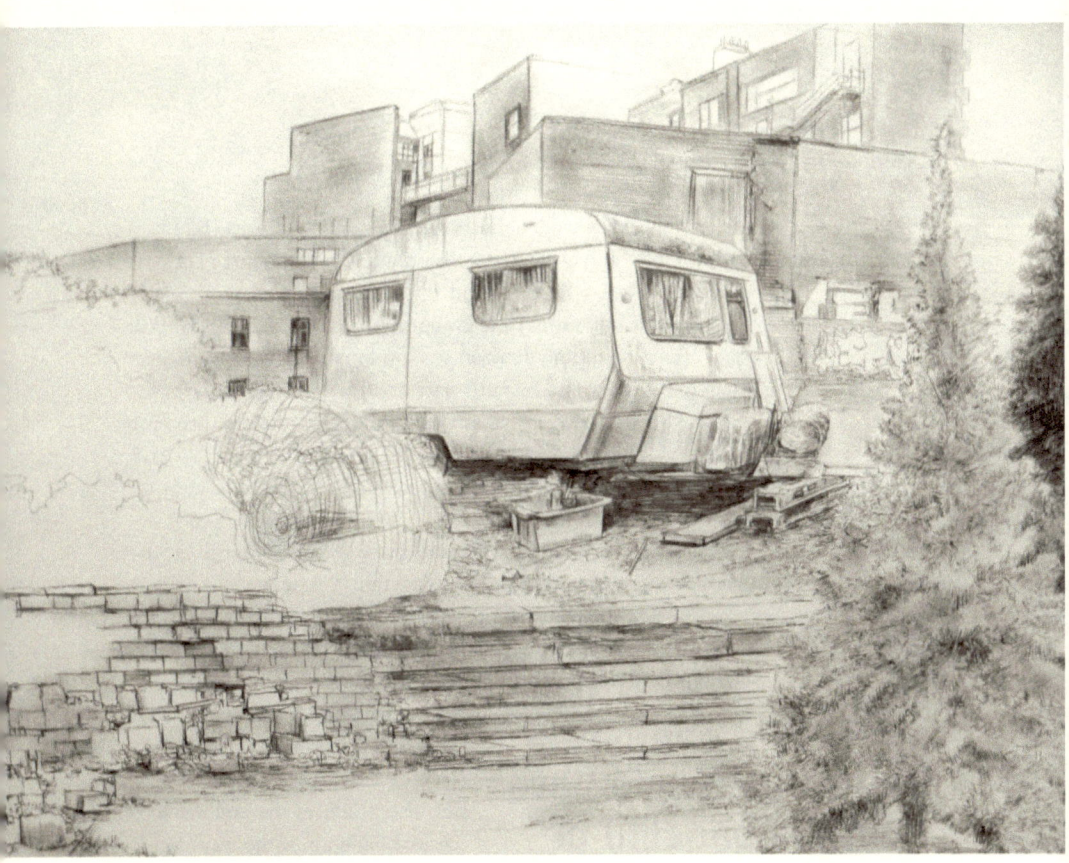

Set We Free: London 2009, Drifting Through the New Builds

Laid out beneath me were the old Camden squatlands...I paused over the shocking montage of Carol Street, a brutal dichotomy, Scritti Politti's Green Gartside reading Derrida on one side and Ian Stuart leafing through Mein Kampf on the other.

– *Savage Messiah*, Issue 8

Hales Gallery is situated within what is now called The Tea Building, a former tea warehouse which is now 'East London's new centre for media and creative industry'. It feels better to be outside the gallery and I walk behind the building to enter what was once called the Nichol (after Nichol St.), but which is more commonly known as the Jago from Arthur Morrison's book *A Child Of the Jago*. Published in 1896, Morrison shone a reforming light on what was then 'the blackest pit in London', infamous for its poverty and its 'howling sea of human wreckage'. Ironically when change came in the form of the Boundary Street Estate two years later, the casual poor, too poverty stricken to afford the new flats, were cast out of the area. I'm not sure we ever remember these previous class cleansings as we walk about in newly gentrified areas like Spitalfields. The cycle just goes on and on, the name 'Tea Building' is mere heritage and not history.

Laura Oldfield Ford is going in (at least) two directions at once. As a long-time denizen of London's squatting, punk and anarchist scenes, she remains part of a trajectory of people who gave a shit back through many and varied tumultuous occasions – Miner's Strike, Class War, Poll Tax riot, J18, Wombles – and into the future. Alongside that more angry and bilious continuum, she has been hard at work – Slade, Royal College of Art, Valerie Beston Award winner and subsequent exhibition at the Marlborough Fine Art Gallery, Mayfair.

Outside the confines of the gallery walls she works with the artist collective We Are Bad, producing uptight and angry denunciations of the London Olympics and frustrating the squeaky clean image of the Games via constant flyposting on the lengthy blue fence of the 2012 exclusion zone. She also produces the zine *Savage Messiah* which in some ways is a lot more confrontational and angry, the imagery less subtle than those on display in the exhibition. It's also a lot more contradictory and its wilder energies seemed harder to appreciate in the deadening silence of the gallery where the first ten issues were wired to a table for your delight.

> **dereliction can be viewed as a safe haven from the banality of yuppie territories**

Full of drawings, photos, stories and rants, memories of drunken nights in Hackney or Elephant, each issue of *Savage Messiah* is situated somewhere in London – a voyage out towards Heathrow, or through the Lea Valley wilderness. Looking through Issue 8, a drift from old King's Cross to Hackney Wick, these are the collisions that hit you – a collage of self-portraits, solid blocks of typewritten text, quotes by Genet or Benjamin or Freud, a skinhead in a defiant pose. 'ASBO DEFIANCE', 'OCCUPY THE OLYMPIC VILLAGE – SQUAT THE YUPPIEDROMES' and an advert for 'Luxury Apartments' defaced with 'Wreck, Loot, Burn'. Photos of the shabby outside of where Rimbaud and Verlaine lived on Royal College Street sit by assorted other fragments: Italian fas-

Chris Jones

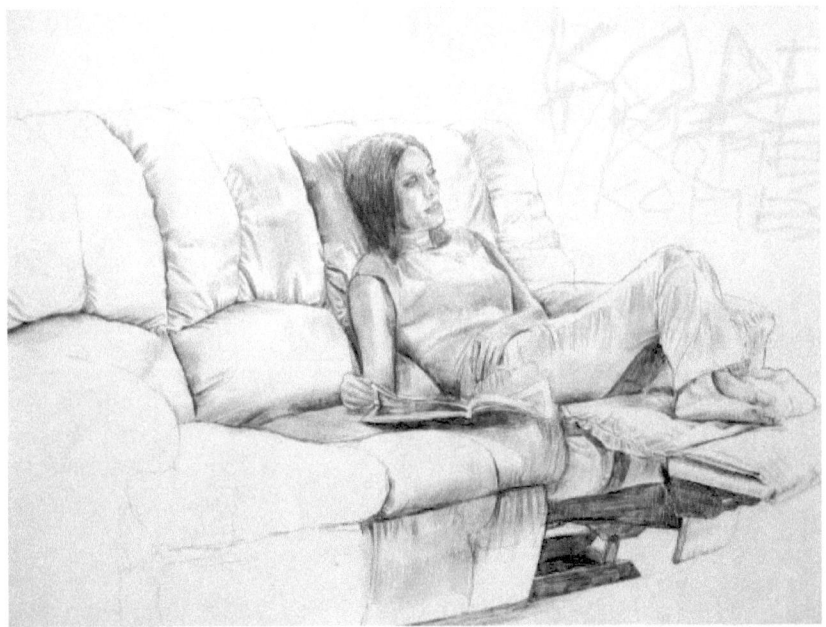

cist graffiti, a junkie injecting into her foot, a skinhead dancing, cops with short shields from the Poll Tax riot, death threats to estate agents and, a *Savage Messiah* favourite, 'ultraviolence'.

This term crops up a lot. Is it a call for proletarian revenge against the toffs and the yuppies? It wouldn't be the most useful analysis of social relations especially in a world of increasingly precarious conditions for many. It's definitely more incisive as parody. The hysteria around the term 'chav' shows that it's common for those with real economic power to still fear the 'savage' nature of the lower orders. Far better for Oldfield Ford to make a bit of cash from selling these drawings to a trendy couple who are happy to objectify the poor with their love of 'council estate chic' than to fall into the overly self-conscious role of the avenging 'savage' from the 'darkest streets'. In my own humble experience, such swagger and anti-intellectualism often work against us in the end, when class politics begins with closure and stereotype wielded against expansion and the self-confidence to be everything instead of nothing.

As a jolly council tenant myself, I'm not so excited by increasing ultraviolence especially when I already feel caught up in a web of disciplinary scenarios. Locally, regeneration schemes turn 'trickle-down' into being pissed on from a great height, and local council biopolitics mean that minor infractions of your tenancy agreement result in Ludivico's Technique in the back room of the housing office.

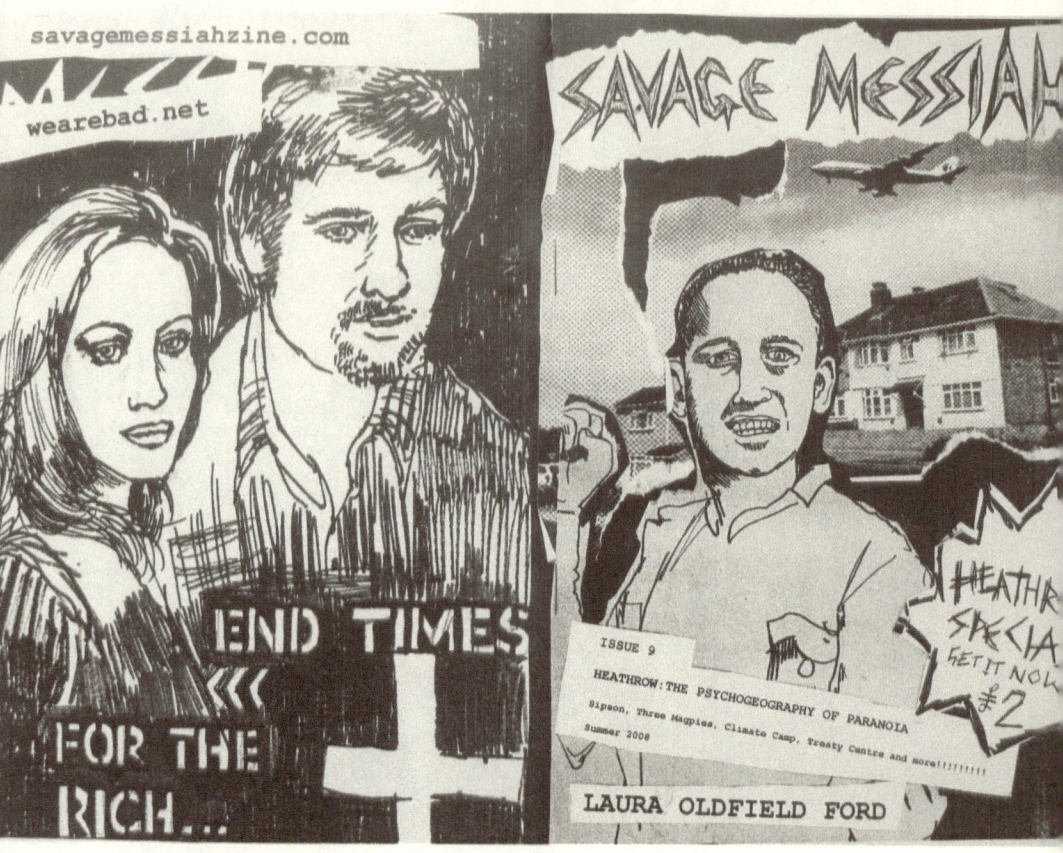

Image: *Savage Messiah*, Issue 9

Between being forced to watch endless films on good behaviour on the one hand, and wannabe gangsters pulling weapons at the slightest agitation on the other, I'm not so big on community atomisation.

For all the desire to expound a utopia for the poor among the debris, and for all the radical history that we can throw at it to inspire ourselves to keep on keeping on, there is also another, more definite non-poetic reality for most of us. Myself, I would prefer my tower block to be a free space of exploration for all, but then I get tired of junkies leaving half-eaten packs of Jammy Dodgers on the landing and prostitutes giving ten quid blow-jobs to Polish labourers outside my front door. These are the ruins we live in then. Ironically it may be the current credit crisis itself that does more to halt regeneration (Olympic or otherwise) than any active class anger, although this may only be their lull before our storm.

Disorientation (Rewind)

I like a great deal of what can be brought to mind from 2013, Drifting Through the Ruins, but then sadly I'm also a bit too long in the tooth to be content with it. I'm not sure where gallery shows sit in relation to all these vital themes – class politics, gentrification, the marvelous in the everyday, exodus. Despite the bile and subtlety, this show sits right at the edge of these themes because it chooses to represent them inside what I consider disputed territory, despite everyone's good intentions.

Ten minutes after leaving the exhibition, I am walking up Great Eastern Street and I feel that familiar disorientation once again. Without irony and with full middle class over-confidence, an expensive clothes shop called A Child of The Jago has filled its window with 'Cashmere Workers Hats' for sale to anyone loaded and clueless enough to sport one. Art may be a useful weapon against this insidious class war, but it will not win the battle as a worm in the bud. Laura Oldfield Ford is right now both inside and outside this process. I prefer the cut and paste immediacy and directness of *Savage Messiah*. Essentially ephemeral, the zine moves round London from hand to hand inspiring a certain righteous gobbyness, remembering but also demanding more rage, more rebellion.

Despite the poignancy and everyday poetry of 2013, Drifting Through the Ruins, I can't separate out these themes from the context of both the constricted, non-feral space of the gallery and its very location. It's hard to avoid the reductionism and the enclosure that goes with these spaces. Hales Gallery is happy to trumpet its recent move into 'the Hoxditch culture project' of the Tea Building. It can also talk about how its 'proper role in the yBa scene of the '90's gives Hales true credibility'. I don't think Oldfield Ford needs her credibility to be reflected back from a commercial art scene. More likely the gallery prefers the reflection her credibility gives to it. It's great what she does, but it lacks a critique of art and the role of the artist. This is the radical position that is spread across the avant-gardes of the 20th century and should not be missed out of any updated project of psychogeography that remains interested in social antagonism. But then these are the complications and contradictions that all politicised artists face at some point and it never pays the rent. What to do?

Info

Laura Oldfield Ford's London 2013, Drifting Through the Ruins was at the Hales Gallery, London, 30 January - 14 March 2009

Savage Messiah, http://savagemessiahzine.com/

Chris Jones <chris56a@yahoo.co.uk> is a prisoner at 56a Infoshop in South London. Visiting times here: http://www.56a.org.uk

CG 2014:
FORMULARY FOR A SKEWED URBANISM

Using the 2014 Commonwealth Games as a launch pad for further analysis, Neil Gray unmasks the dehumanising and exploitative realities behind the logic of urban regeneration strategies in Glasgow

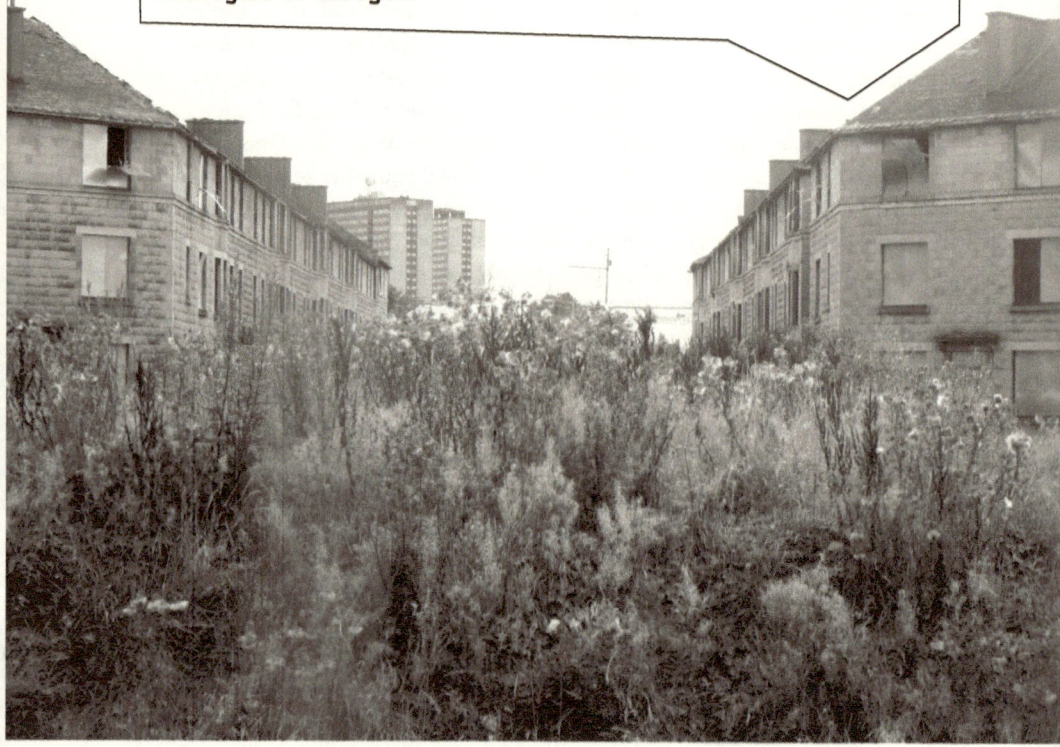

All images courtesy of the author. Image: Tenements in Oatlands, part of the Clyde Gateway area

Neil Gray

The Games offer our country a chance to advertise to a global audience of over 1 billion people. Glasgow is an incredible city and Scotland is an unforgettable country. The more people who get the chance to see this the more we can grow in the future.

– Glasgow 2014 Ltd[1]

On Friday 9 November 2007, amidst saltire waving hysteria, the General Assembly of the Commonwealth Games Federation chose Glasgow as the host city for the 2014 Games. The Deputy First Minister, Nicola Sturgeon, of the Scottish National Party (SNP) captured the zeitgeist as imagineered by the 'Team Glasgow' neoliberal consensus, albeit tinted with her own nationalist hue: 'This will bring a host of benefits to Glasgow and Scotland, including everything from regeneration, job creation, inward investment and just a huge pride in being Scottish'.[2] Between the often convergent claims of 'Team Glasgow' and 'Team Scotland', the Games have been declared an outstanding success five years before they've even begun. This should come as no surprise; the rolling out of Bread and Circuses is about the most coherent political strategy available to city regions under the *external coercive power* of neoliberal inter-urban competition. As a senior planning officer from Manchester Local Authority explained of Manchester's Commonwealth Games in 2002:

> Once you adopt a strategy of pursuing sports-led urban regeneration, politically it is very difficult for it to be allowed to fail [sic], so what happens is that it gets declared a success, really irrespective of what happens on the ground.[3]

Mega sporting events are now routinely staged in 'run down' areas of host cities and mobilised as both alibi and instrument for the 'regeneration' of declining urban regions. The Commonwealth Games 2002 in Manchester and London's Olympic Games 2012 are paradigmatic examples in the British context. The notoriously deprived East End of Glasgow is the latest 'urban frontier' to fall prey to this transparent property development strategy. The Games will take place over 12 days from 23 July to 3 August, 2014, and Glasgow 2014 Ltd – comprising the Scottish Government, Glasgow City Council and the Commonwealth Games Council for Scotland – is the organisation charged with overseeing the management of the event. While the figure routinely reported in the press is a downsized £350 million, in fact, an estimated £2 billion of public money will go towards the costs, including the construction of a new Indoor Sports Arena, a Velodrome, and an Athletes' Village, alongside supporting physical infrastructure.[4] City boosters, meanwhile, claim the event will stimulate the building of 1,000 new homes, and create 1,000 permanent new jobs.

Glasgow's Commonwealth Games 2014 (CG 2014) cannot be adequately understood without grasping its relationship to wider waterfront regeneration ambitions along the £5.6 billion, 13-mile Clyde river corridor. The 'return to the river' along a post-industrial, disinvestment valley around the city centre's financial district, is seen as a key priority in transforming Glasgow's image from recalcitrant Red Clydeside to, newly branded, 'Glasgow: Scotland with Style'. While much of this development has taken place around geographically circumscribed areas in already developed city spaces, the Clyde Gateway Initiative in the East End, at 2,070 acres in size, is a potent 'new urban frontier' – a potential land-grab of enormous proportions.

a 'discourse of decline' represents 'renewal' and 'regeneration' as both natural and irresistible

To this end, the initiative aims to oversee £1.5 billion of private sector investment, 400,000 square metres of business space, 21,000 new jobs, 10,000 new houses, and a population increase of 20,000 people over 20 years. The rhetoric of 'regeneration' in the area has pitilessly converged on 'blight' and the (un)productivity of local land and labour as a justification for wholesale urban renewal.[5] But the paternalist discourses attached to the regeneration script mask pernicious undercurrents: territorial stigmatisation, an ideological drive towards 'flexible' labour markets and a continuing thrust towards private property development.

Imitation of Life

In the mid-'90s, Neil Smith pursued an argument that Frederick Turner's influential essay 'The Significance of the Frontier in American History' (1893) had crucial resonance for critics of urban gentrification. Put simply, for Turner, the Western frontier was envisioned as 'the meeting point between savagery and civilization'; each wave westward in the conquest of people and nature contributed to new enclosures of land, space and labour, part of a wider mission to civilize an unruly and uncooperative landscape. By the latter part of the American 20th century, Smith contends, Turner's imagery of wilderness and the frontier was being applied 'less to the plains, mountains and forests of the West [...] and more to cities back East'. In the modern reconfiguration of frontier lines, many slums in major US cities were demarcated as 'urban wilderness' in order to legitimise their removal through slum clearance and urban renewal.[6]

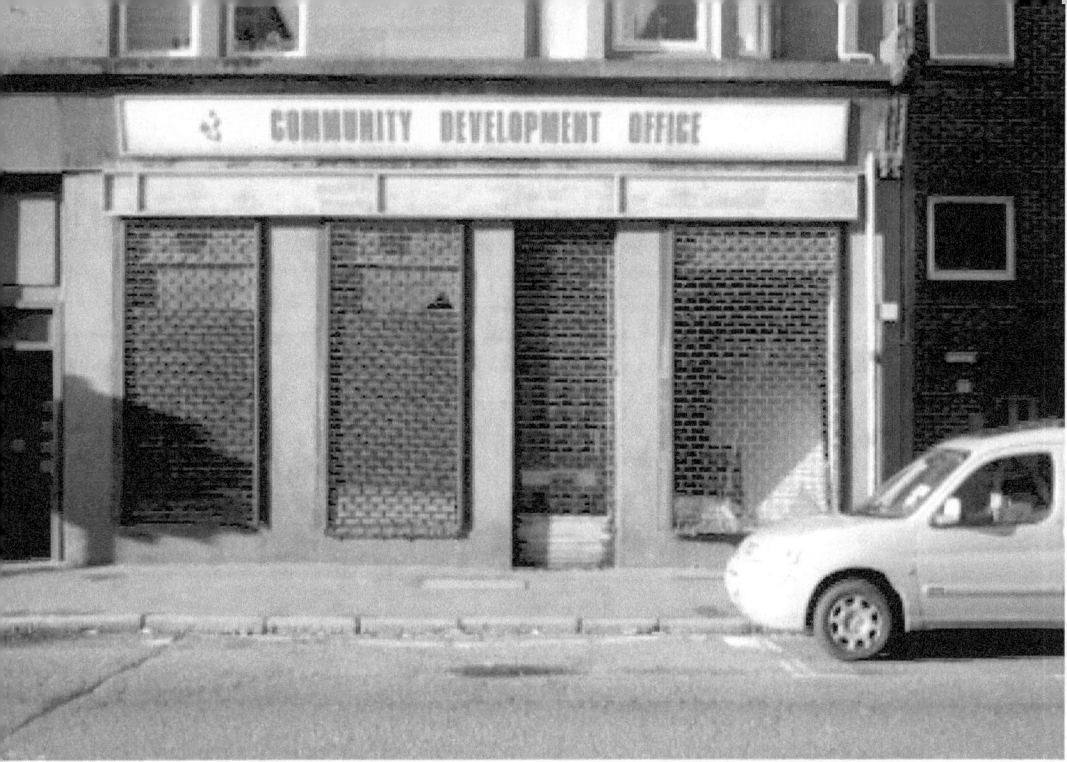

Image: Community Development Office, Bridgeton, a key area in the Clyde Gateway proposals

By the 1970s, this 'discourse of decline' was accompanied by boosterist discourses of urban development through gentrification. By the 1980s these entrepreneurial discourses had intensified – the 'urban jungle' would be put to the sword by a new breed of urban hero. In the Reagan era, 'urban pioneers', 'urban homesteaders' and 'urban cowboys' were represented as the new 'folk heroes of the urban frontier'. For Smith, the important conclusion to be drawn from these frontier discourses, is that they attempt to 'rationalise and legitimate a process of conquest, whether in the 18th and 19th century American West, or in the late 20th century inner city'. Like a domestic form of Orientalism, the frontier motif encapsulates a host of accumulated symbolic meanings with embedded uneven power relations, including:

> the social differences between 'us' and 'them', the historical difference between past and future, and the economic difference between existing market and profitable opportunity.[7]

The imagery of conquest and expansion associated with the frontier motif has taken a kicking in the current economic crisis, yet frontier imagery persists.[8] One method for dealing with capitalism's problem of inherently crisis ridden over-accumulation is *devaluation,* and the frontier analogy – associated as it is with notions of blight, decay and wilderness – can help provide a 'convenient incantation' for development strategies. If the frontier doesn't exist, the state will attempt to create it. The state's willingness to subject its property and land base to market rule, and its simultaneous requirement to manage and disperse native populations, accounts for the zeal with which it stigmatises certain people and certain places. By targeting urban working class populations as 'less than social', and the 'frontier' area as 'not yet socially inhabited', the state attempts to juggle the political imperative of administering potentially recalcitrant local populations, with the entrepreneurial task of maintaining or creating the conditions for profitable capitalist investment. Glasgow East – with its large tracts of derelict land and deeply impoverished population – lends itself to this 'discourse of decline'; a discourse that in turn represents 'renewal' and 'regeneration' as both natural and irresistible.[9]

A Time to Love and a Time to Die

The July 2008 UK parliamentary by-election in Glasgow East became a crucible for territorial stigmatisation. Traditionally a staunch Labour constituency, Glasgow East was typically represented as a potent symbol of what the Conservative Party have called Britain's 'broken society', and thus an emblem of Labour's political failure. A previous visit by Ian Duncan Smith to the East End 'inspired' the formation of Tory think tank the Centre for Social Justice, which developed the extremely derogatory term 'Shettlestone man' to profile male individuals who make up the largest constituency in Glasgow's East End: 'This individual has a low life expectancy. He lives in social housing, drug and alcohol abuse play an important part in his life and he is always out of work'.[10] Disavowing any consideration of the *historical* and *political* construction of urban space under successive capitalist regimes, press and politicians instead embarked on an overwhelmingly negative assault on Glasgow East and its population. The area was variously described as 'the hardest, poorest place in Britain'; 'a ghetto ringed by some of the saddest statistics in Britain'; a 'hell-hole'; a 'welfare ghetto', and an 'invisible' no-go zone. Worse, in a cruel conflation of place and people, the residents of the Glasgow East constituency – allegedly a 'social disaster' where the 'law of the jungle' prevails – were frequently described with unapologetic class hatred: 'The people do not look good here. Often it is difficult to tell men from women, old men from older men [...] the locals have the blotchy pallor of cave-dwelling consumptives'.[11]

despite its failure, the 'moral capital' attached to business activity is at Victorian levels

The not so hidden discourse behind all this vile rhetoric was an attack on welfare and the so-called 'dependency culture'. David Cameron claimed the area was

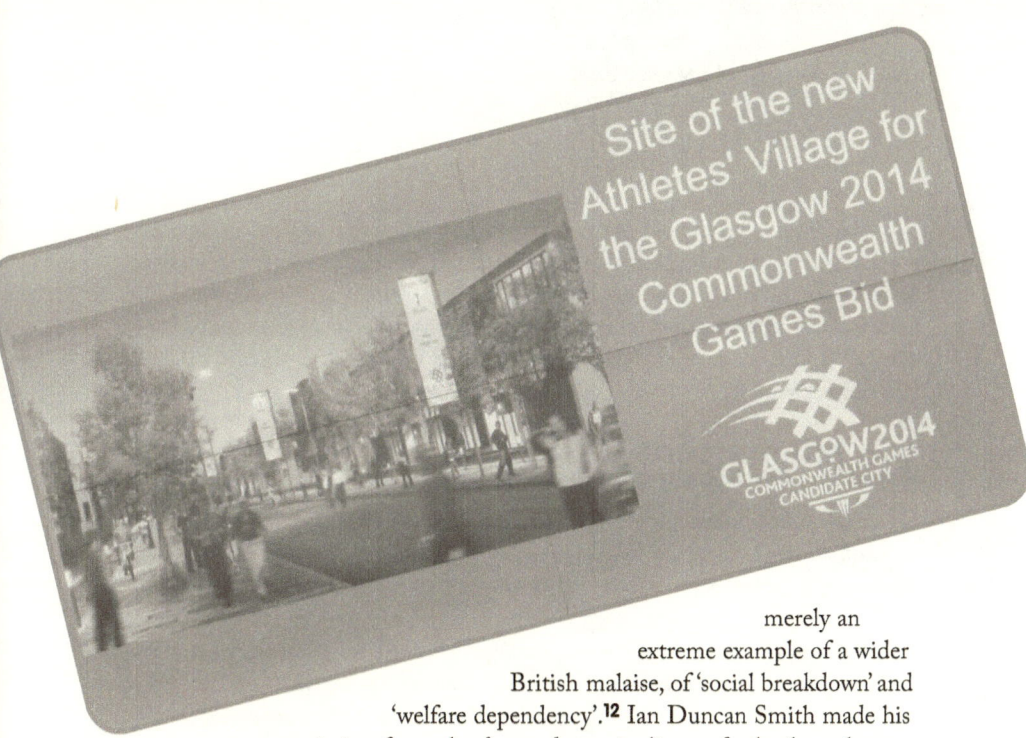

merely an extreme example of a wider British malaise, of 'social breakdown' and 'welfare dependency'.[12] Ian Duncan Smith made his prescription for such advanced marginality perfectly clear: the people of Glasgow East need to get the 'work habit'. The press lustily joined in with this chorus, fancifully claiming that Glasgow East, a post-industrial area hammered by decades of labour exploitation, sustained disinvestment, de-skilling, slum clearance, depopulation and *retrenched* welfare services, was serviced by 'epic amounts of public money'. Glasgow East and its 'welfare prisoners' were represented as 'a hideous, costly social experiment gone wrong', that cost 'billions to achieve'. Isn't Glasgow East proof, one journalist asked, of 'just how utterly poisonous that sort of thing is?'[13]

James Purnell's Welfare Reform White Paper proposals show, unsurprisingly, that Labour is entirely in concord with the Tory right. Smith's Centre for Social Justice boasts of how many of its recommendations are now advocated by Purnell on its website. David Freud, the *investment banker* who was Purnell's key advisor on welfare reform, meanwhile, has recently switched allegiance back to the Tories after previous tugs of war between the parties to enlist his advice on involving the private sector and voluntary groups in forcing the 'jobless' into the labour marketplace.[14] Purnell's proposed welfare reforms would see a Flexible New Deal (FND) that would refer claimants who'd been out of work for 12 months to *incentivised* private or third sector contractors, who *would be paid by results to find them work*. Added to this, Purnell proposes cuts in income support for single parents, cuts in incapacity benefit for disabled people, increased sanctions (benefit cuts) for benefit claimants, and the introduction of US-style workfare pilot schemes. The 'moral capital' attached to business activity, despite evidence of its failure all around, is at

Image: A sign at Dalmarnock, site of the Commonwealth Games Village

Victorian levels, and the use of the private and voluntary sectors is 'business as usual' for the Department for Work and Pensions. With £2 billion worth of contracts up for grabs over the next five years, the poverty industry is a growing business. Glasgow East, with its allegedly 'unproductive' population (unproductive for whom?) is a key node of struggle in the proposed UK-wide transition from 'welfare to workfare'.

The Magnificent Obsession

What we want to do is give people the chance to get back into the labour market, that's my understanding of a successful growing economy.
— Ian Manson, Clyde Gateway URC[15]

As far as I am concerned, business is Santa Claus, but there is still a passive attitude that sees it as a necessary evil rather than something that is fundamentally good.
— Richard Cairns, Glasgow Chamber of Commerce[16]

Glasgow City Council has set itself stringent targets to 'overcome' what it sees as a GDP productivity gap. The City's Business Forum aspires to increase productivity by 43 percent and add another 50,000 to the employment register. Key strategy documents stress the need for Glasgow to move up the 'value chain' in employment and employability terms, yet the Clyde Gateway Initiative and CG 2014 are unable to offer anything more than the same old low-paid, insecure service sector jobs. Nine out of ten jobs in Glasgow are in the service sector with even the City Council acknowledging that these jobs are in 'lower paid and lower skilled services'.[17] While the city dutifully adheres to the discredited creative economy thesis, it refuses to face up to the creative economy's mirror image: that of it's supporting infrastructure; an expanding reserve army of low paid, insecure service workers.[18] In this context, the promise of the Commonwealth Games organisers to provide up to 1,000 jobs, and Clyde Gateway Initiative's promise of 21,000 jobs looks more like a threat rather than an opportunity for the unemployed.

Ignoring the grim realities of wage labour for the post-industrial working class – who are increasingly forced to choose between declining welfare levels in real terms and residualised, insecure and alienating, low paid service work – the employment debate is increasingly framed by the burgeoning third sector and private agencies in terms of aspiration, individual responsibility, volunteering and training geared towards the needs of employers. Yet even these limited 'opportunities' are unrealisable. While the City Council has promised 5,000 apprenticeship places for the Games, including 2,000 apprenticeships in construction, the collapse of the housing market and the deficit in existing training capacity mark these claims out as 'hollow rhetoric' to sell the Games.[19] In 2007, for only 75 posts in Glasgow City Council's four year apprenticeship scheme ('City Building'), there were 2,000 applications. A gift to employers, but a significant deficit for young people. Meanwhile, less than 200 jobs were created in Glasgow's construction sector between 2001 and 2006 (in notably more auspicious times for development strategies), which hardly bodes well for aspirant trainees in the current economic climate. With the absence of worthwhile training opportunities many 'deprived' young people are being corralled into volunteering projects: the Commonwealth Games, set to be run by a massive reserve army of 12-15,000 *unpaid* voluntary 'workers', can be seen as a key experiment for the inculcation of the 'work ethic' through unpaid labour.

> **Welfare reform is the stick that beats the subject into the alienating embrace of wage labour**

None of these initiatives can mask the fact that the wage labour relation itself has become a rampant source of fragmentation and desocialisation: the burgeoning ranks of the 'working poor' are testament to that. In the 'post-Fordist' era, on the sharp edges of the (un)employment sphere, wage labour is a source of built-in precarity at the bottom of the class structure. In this context, it is a delusion to think that pulling people into the labour market will reduce poverty, as the market is dominated by bottom-end service jobs. The dull, market-led compulsions of current welfare reform proposals, meanwhile, can be seen as both an attack on the unemployed and an attack on workers in jobs, as competition will exert even more downward pressure on wages and conditions. Welfare reform throws the subject into the alienating embrace of capital and wage labour; thus the intention is to create a labour pool, 'that belongs to capital quite as absolutely as if the latter had been bred at its own cost.'[20]

Tarnished Angels

> *We are aware the Government wants to grow Scotland's economy and to do that, it needs to bring all the land back into economic use.*
>
> – Ian Manson, Clyde Gateway URC[21]

As noted earlier, territorial stigmatisation is particularly potent at the level of urban 'regeneration'. The language deployed intentionally obfuscates fundamental questions over the exploitative production of space historically, and displaces questions of culpability and collective responsibility away from the state and business sectors. Counter to this a-historicism, Neil Smith's analysis of gentrification and the new urban frontier relies on an understanding of the historical movement and control of capital in the built environment. As a key model for the post-industrial city, of which Glasgow is a prime exemplar, Smith usefully returned to Homer Hoyt's analysis of Chicago and other industrialised North American cities in the 1930s.[22] The classical 19th century 'conical form' found by Hoyt provides a useful, if incomplete, model for understanding the economic imperative behind the back-to-the-city movement by capital in the Glasgow context.

In Hoyt's model, land values displayed a high value peak at the urban centre with a declining value gradient on all sides towards the periphery. As industry declined, or was abandoned, the long term investments in capital and land associated with previous industrial development – and associated renewal and redevelopment costs – created barriers to capital's ability to expand *profitably* in inner city areas. An important exception to this model was typically made in the Central Business District (CBD) where investment and maintenance tended to continue. Unwilling to invest in central urban industrial locations, capital leapt out to the edge of the land value cone, on the city fringes, where spatial expansion was both possible and cheap. As a result, investment around inner city areas was curtailed and land values fell, often drastically, relative to the CBD and the suburbs. By the late 1920s Hoyt could identify a 'valley in the land value curve', between the CBD and the outer areas of the city. This area Hoyt termed the 'disinvestment valley'.[23]

the cause of urban poverty is typically mistaken for the remedy

Central to Smith's argument is that the capital devalorisation and urban deterioration so typical of Hoyt's thesis is 'a strictly logical, "rational" outcome of the operation of the land and housing markets.'[24] Disinvestment in one area allows capital to be profitably invested elsewhere. Meanwhile, the devalorisation or 'blight' associated with disinvestment produces the objective economic conditions that eventually make capital revaluation, or gentrification, a 'profitable' market response. In the 'disinvestment valley', Smith contends, arise the conditions that make possible the new urban frontiers of gentrification. The fate of those residents who have suffered for decades in this tortuously slow cycle of creative destruction has been to endure the wrath of territorial stigmatisation, deflecting critical enquiry away from previous policy and economic decisions, and providing a convenient alibi for 'regeneration' and working class displacement.

Neil Gray

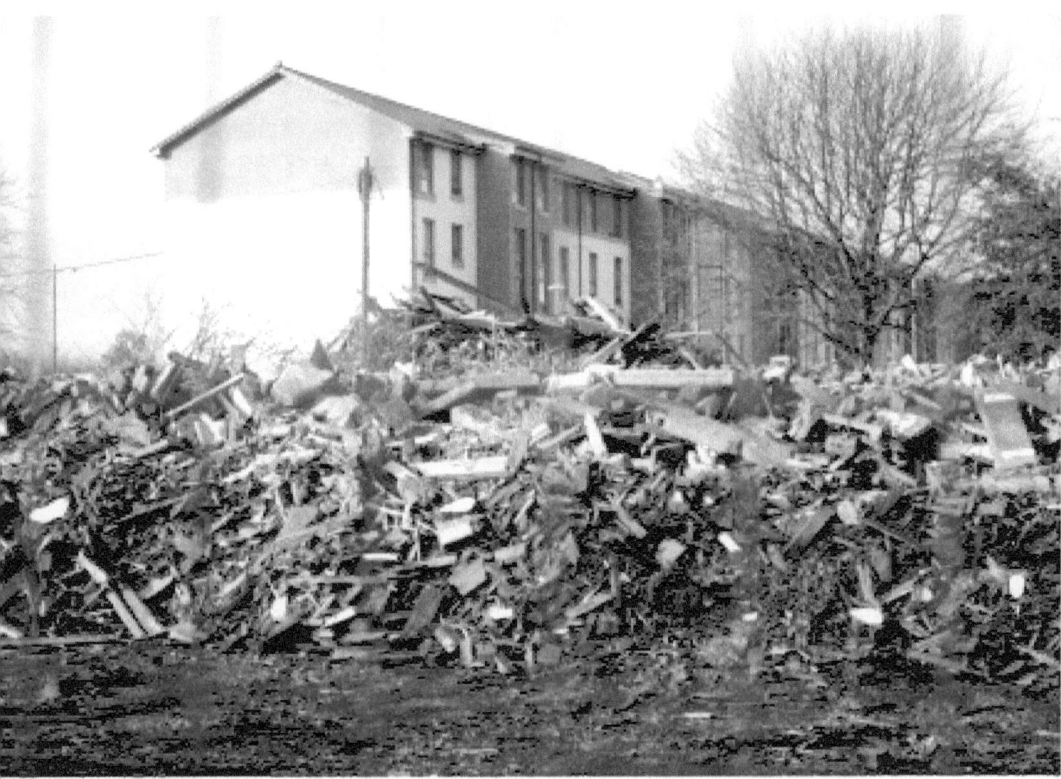

Image: Demolition in Oatlands

Written in the Wind

In the context of Glasgow's urban topography, the Hoyt thesis should resonate with anyone who has ever walked through the 'doughnut' of disinvestment around the city's former riverside industrial core.[25] As inner city development spreads out from the city centre, development plans for the de-industrialised areas in the east of Glasgow can be seen in the context of a developing 'rent gap' related to property expansion along the Clyde river corridor.[26] As Smith suggests, these developments have less to do with the 'hidden hand' of the market, and more to do with a series of identifiable private and public investment decisions at the service of the market. The high incidence of derelict land in the east of Glasgow, the socialisation of risks and costs in its remediation (essentially 'geo-bribes' to lubricate capitalist investment), and its proximity to property development in the Merchant City and the near east of the city, make the area a prime, and *primed*, target for property developers.

Scottish government figures for 2006 claim the percentage of people living within 500 meters of any derelict site in Shettleston was 79.1 percent, while in nearby Calton the figure rises to a staggering 99.4 percent.[27] These conditions reflect the successive failures of capitalist administrations to deal with some of the worst urban poverty in Britain. Yet the *cause* of these problems is typically mistaken. Glasgow City Council, a consummate agent of the neoliberal project, makes clear in the 'City Plan' that vacant and derelict land alongside the Clyde development corridor represents 'a significant development opportunity if appropriately remediated and marketed'. The Council will 'continue to use its property and land portfolio to unlock development opportunities and stimulate the market.'[28] To this end a £13.5 million Vacant and Derelict Land Fund programme – part of a £50 million 'Better Glasgow' 'regeneration' strategy – has been set up citywide to prime Council-owned sites this year for use by private developers and businesses. Moreover, regarding CG 2014, City Council leader Stephen Purcell has maintained that the Council will sell 'surplus property and land' to meet the costs of hosting the event, while a Council spokesman said that land and property worth 'hundreds of millions of pounds' was available for sale.[29] The Council reportedly wants to 'transfer' '56 surplus sites' to a new joint venture as part of the funding for the event.[30]

In case the business community is concerned that the current recession will curtail some of the Council's traditional largesse, Mr. Purcell has been at pains to assure it of the continuing loyalty of 'Team Glasgow' to the crisis-ridden neoliberal project. Alongside promises of continued 'flexibility' in terms of the disposal of publicly held land and assets, he told assembled guests at the recent State of the City Economy conference that Glasgow would remain 'open for business':

> My main priority is helping business in the city through the economic difficulties ahead [...]. The first thing that all public bodies, including my own Council, must do is to examine where we can help business by being more flexible and willing to do things differently. This is no time for unnecessary rules and processes; this is a time to do everything we can to help.[31]

All that Heaven Allows, or, All I (We) Desire?[32]

The 'overarching purpose' of the Commonwealth Games 2014, the Games Interim Bid Legacy document reminds us, is to increase 'sustainable economic growth'. This aim reflects the normalisation within regeneration discourses of the, now severely impaired, 'moral capital' of private enterprise. Competition and the pursuit of wealth over recent decades have been routinely valorised as ends in themselves, with 'trickle-down' effects supposedly guaranteed. Yet inter-urban competition, expressed through symbolic mega-events like the Commonwealth Games and large-scale regeneration projects, is, and always was, in reality a zero-sum game of 'trickle up': the 'external coercive power' of inter-city competition creates an accelerating race to the bottom, via the disposal of public assets and the exploitation of the labour force, in order to foster the 'good business climate' which other cities also strive to achieve – in this way every city is merely brought further into line with the discipline and logic of capitalist development.[33] The end results of this competitive schema are a banality:

> Competition, as Marx long ago observed, always tends towards monopoly (or oligopoly) simply because the survival of the fittest in the war of all against all eliminates the weaker firms [...] [T]he fiercer the competition the faster the trend towards oligopoly if not monopoly.[34]

If experience shows us that neoliberal policies everywhere produce more poverty, more insecurity, and higher socio-economic inequality, the retreat to nostalgic forms of Fordist regulation is not a viable alternative in itself. In the face of de-socialised wage labour, precaritisation and generalised working poverty, Keynesian or 'social democratic' modes of state intervention are likely to 'stall, disappoint and eventually discredit themselves, paving the way for a further expansion of market rule.'[35] In order to provoke a 'crisis of legitimacy' in both neoliberal and Keynesian orthodoxy, attention might be better directed towards a more radical critique of the essential precondition of capitalist development – wage labour itself. What distinguishes radical practice from leftist, social democratic practice, after all, is precisely the difference between the struggle to *mitigate* alienation in the workplace through various measures, and the struggle to *supersede* alienation and the wage labour relation per se. However, most people still have to work, however precariously, and the workplace will continue to be the key site of *solidarity* in the struggle against capitalism. In order to avoid the perennial dichotomy of 'reform' and 'revolution', those involved in the struggle against capitalism might do well to engage the problem dialectically and *agitate* the once productive tension between *minimum* demands (for 'fair' wage levels, working conditions, environmental conditions etc.) and *maximum* demands (the overthrow of capitalism and the alienating wage labour contract).

The scale of developments in the East End of Glasgow, the symbolic weight attached to the wage labour contract through welfare reform, the inevitable increase in overall city unemployment, and the structural crisis of the capitalist economy mark it out as a key site from which to contest the fundamental contradictions of capitalism. A critical feature of this struggle should be to unmask the instrumental strategies behind the stigmatisation of the local population, and reject outright the 'remedy' of workfare. Meanwhile, as the recession tightens its grip the spectre of Keynes is bound to loom large. But a challenge to capitalism which is not just limited to the 'disjunctive synthesis' of representative democracy will have to question the tacit consensus behind the ownership and management of productive forces.[36] Confining struggle to the sphere of distribution – in the probable form of banal service jobs, exploitative commodities, jerry-built housing, and the dreary spectacle of Bread and Circuses – won't be, and never has been, enough.

Footnotes

[1] See http://www.glasgow2014.com/

[2] See http://news.bbc.co.uk/1/hi/scotland/glasgow_and_west/7086680.stm

[3] Larissa E. Davies, 'Sport and the Local Economy: The Effects of Stadia Development on the Commercial Property Market', *Local Economy*, Vol.23, No.1, February 2008, pp.31-46.

[4] Interim Games Legacy Plan for Scotland, 2008.

[5] "The need for such an initiative is evident from the concentration of economic, social and physical deprivation found in the area. It suffers from high levels of unemployment and low levels of economic activity; from social deprivation and poor health; and, from a concentration of derelict and contaminated land that blights the physical environment." See http://www.southlanarkshire.gov.uk/coins/commpdfs/public/1213.pdf

[6] Neil Smith, *The New Urban Frontier: Gentrification and the Revanchist City*, Routledge, 1996, Preface. See also, Neil Gray, 'The Clyde Gateway: A New Urban Frontier', *Variant Magazine*, No. 33, Winter 2008, http://www.variant.randomstate.org/33texts/3_V33gray.html

[7] Ibid.

[8] For news of a very significant developer collapse in Edinburgh – and of the long resistance to said developers – see http://www.eh8.org.uk/

[9] See Rachel Weber, 'Extracting Value From the City: Neoliberalism and Urban Redevelopment', in, *Spaces of Neoliberalism: Urban Restructuring in America and Western Europe*, Blackwell Publishing, 2002.

[10] See Ian Duncan Smith, 'Living, and dying on welfare in Glasgow East', *The Telegraph*, 24 December 2008, http://www.telegraph.co.uk/comment/3560470/Living,-and-dying-on-welfare-in-Glasgow-East.html

[11] All quotes from research compiled by Gerry Mooney, 'The Politics of Poverty and the Politics of Place: Reporting Glasgow East', forthcoming in *Social Policy and Society*, late 2009.

[12] In Shettleston, the Scottish parliamentary constituency at the heart of Glasgow East, 34.9 percent of the population are 'income deprived', with 30.1 precent 'employment deprived', See http://www.sns.gov.uk/Reports/Report.aspx?ReportId=2&AreaTypeId=SP&AreaId=44 . In Dalmarnock, the site of the Games Village and National Indoor Stadium, in 2000, 48.9 percent of the population were incapacity benefit claimants, and 61.6 percent considered 'economically inactive', Public Health Information manager, NHS Scotland.

[13] Ian Duncan Smith, op. cit.

14 See http://www.guardian.co.uk/politics/2009/feb/16/freud-defects-labour-conservatives
15 Iain Lundy, 'My vision will bring the wow factor to East End', *The Evening Times*, 9 May 2008.
16 See http://www.sundayherald.com/business/businessnews/display.var.2392683.0.full_of_east_end_promise.php
17 The Glasgow City Council Plan, 2008-2011.
18 See Neil Gray, 'The Merchant City: An "Arts Led Property Strategy"', *Variant Magazine*, No. 34, Spring 2009.
19 Andy Cumbers, Gesa Helms, Marilyn Keenan, 'Beyond Aspiration: young people and decent work in the deindustrialised city', preliminary paper given at the Alternative Economic Strategy Seminar, 'Good jobs, nae jobs, bad jobs: Skills, workfare and struggles over work', Glasgow, February 2009.
20 Karl Marx, 'The General Law of Capitalist Accumulation', *Selected Writings*, Oxford University Press, 2001, p.517.
21 Iain Lundy, 'My Vision will bring the wow factor to East End', *The Evening Times*, 9 May 2008.
22 Over the last ten years, the manufacturing sector in Glasgow has reduced by 31 percent to its current level of 6 percent. Meanwhile, the service sector now accounts for nine out of ten jobs in Glasgow.
23 Neil Smith, *The New Urban Frontier: Gentrification and the Revanchist City*, Routledge, 1996, pp.51-74.
24 Ibid.
25 See Anthony Iles, 'Drifting Away From Reformist Politics in Glasgow', http://www.metamute.org/en/drifting_away_from_reformist_politics_in_glasgow
26 'The rent gap is the disparity between the potential ground rent level and the actual ground rent level capitalized under the present land use', Neil Smith, op. cit.

27 See http://www.sns.gov.uk/Reports/Report.aspx?ReportId=2&AreaTypeId=SP&AreaId=44
28 See http://www.glasgow.gov.uk/en/Business/City+Plan/
29 See http://www.eveningtimes.co.uk/misc/print.php?artid=1306800
30 See http://scotlandonsunday.scotsman.com/business/Glasgow-will-struggle-to-sell.4260309.jp
31 See http://www.glasgow.gov.uk/en/News/Archives/2008/November/stateofthecityeconomy2008.htm
32 The section titles are from the films of Douglas Sirk – who charted the disjuncture between the surface vigour of capitalism (now imploding) and all that it obfuscates with glorious clarity.
33 David Harvey, 'From Managerialism to Entrepreneurialism: The Transformation in Urban Governance in late Capitalism', *Geografiska Annaller*, Vol.71, No.1, pp.3-17.
34 David Harvey, 'The Art of Rent: Globalisation, Monopoly, and the Commodification of Culture', 2002, http://socialistregister.com/recent/2002/harvey2002
35 Lois Wacquant, *Urban Outcasts: A Comparative Sociology of Advanced Marginality*, Polity Press, 2008, p.252.
36 Michael Hardt and Antonio Negri, *Multitude*, Penguin Books, 2005, p.241.

Neil Gray <neilgray00@hotmail.com> is a writer and film-maker based in Glasgow

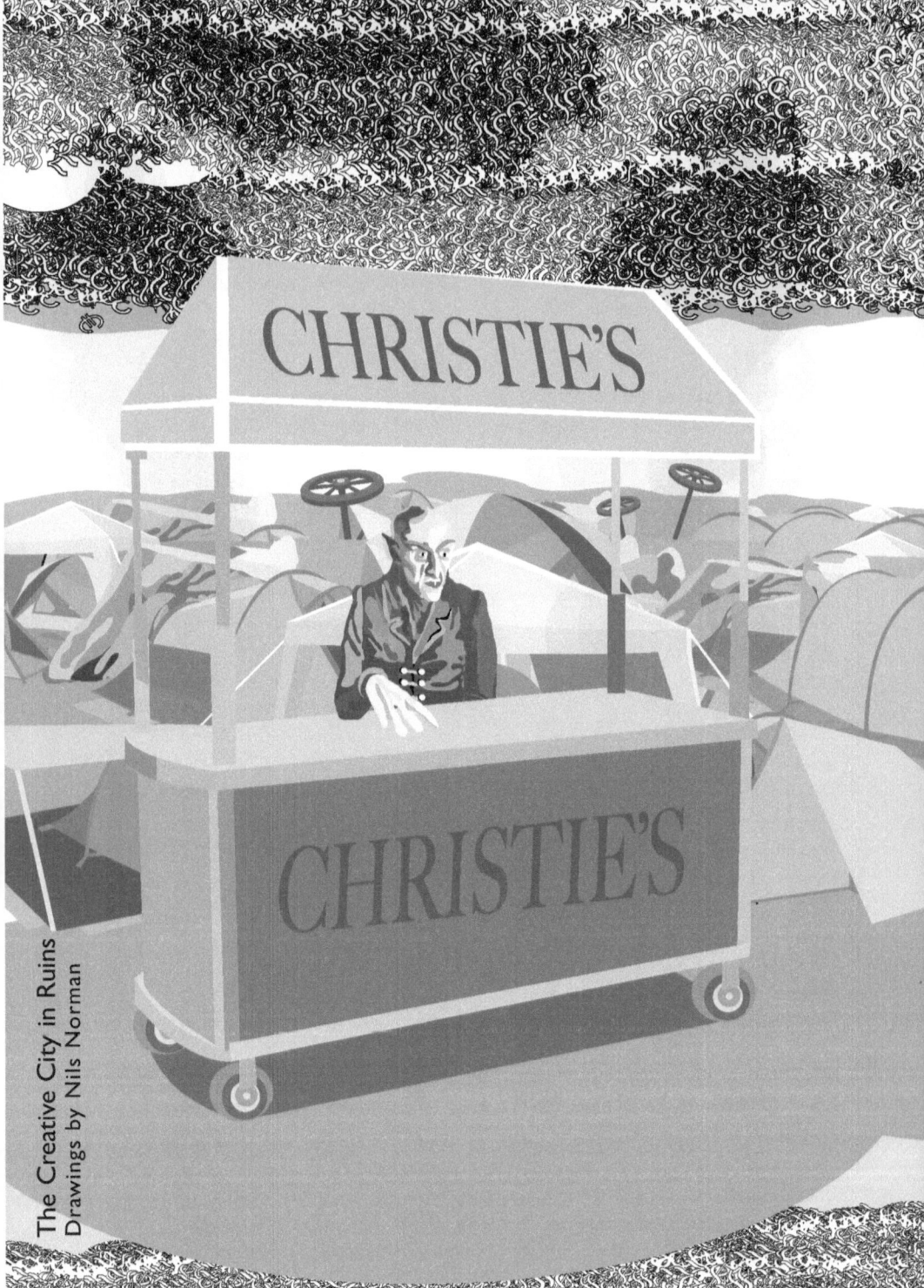

The Creative City in Ruins
Drawings by Nils Norman

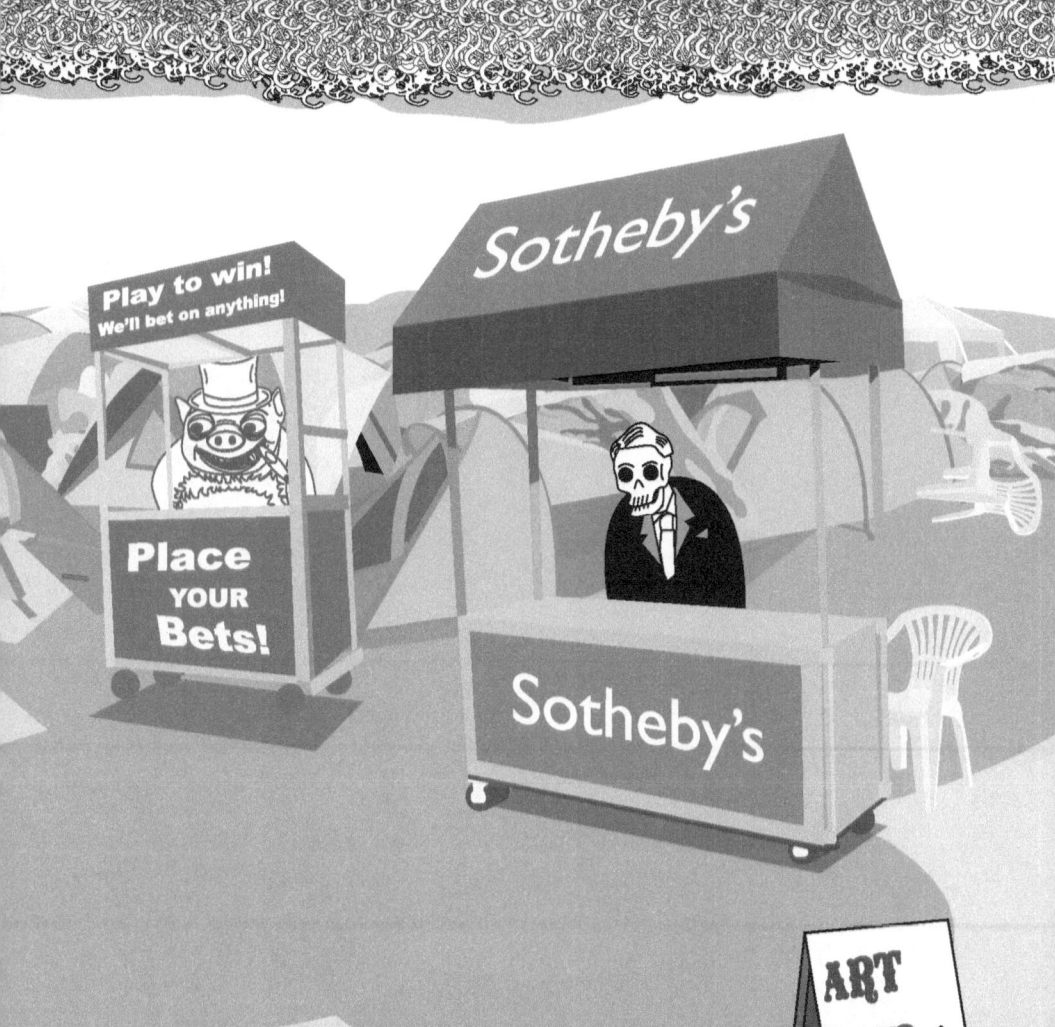

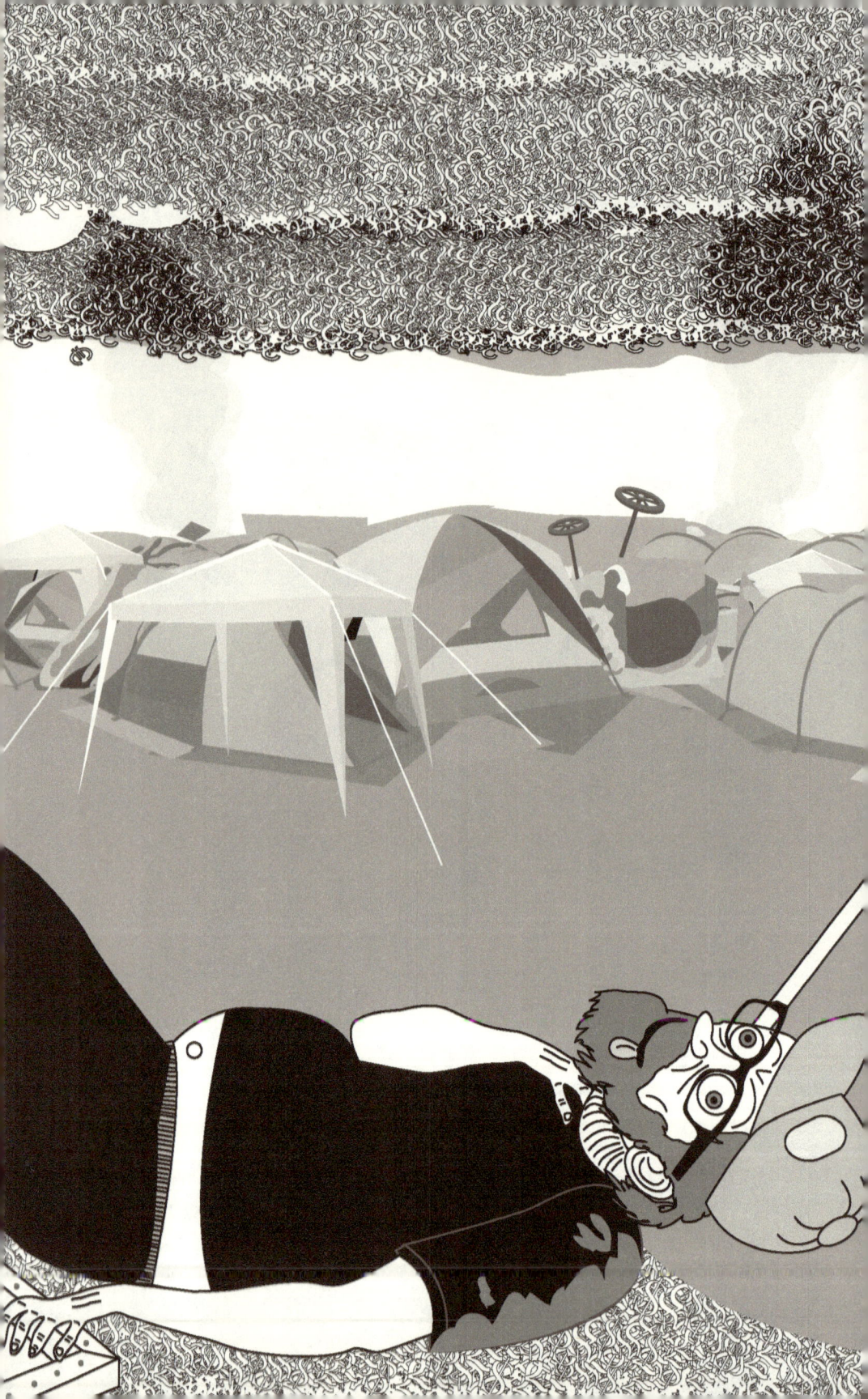

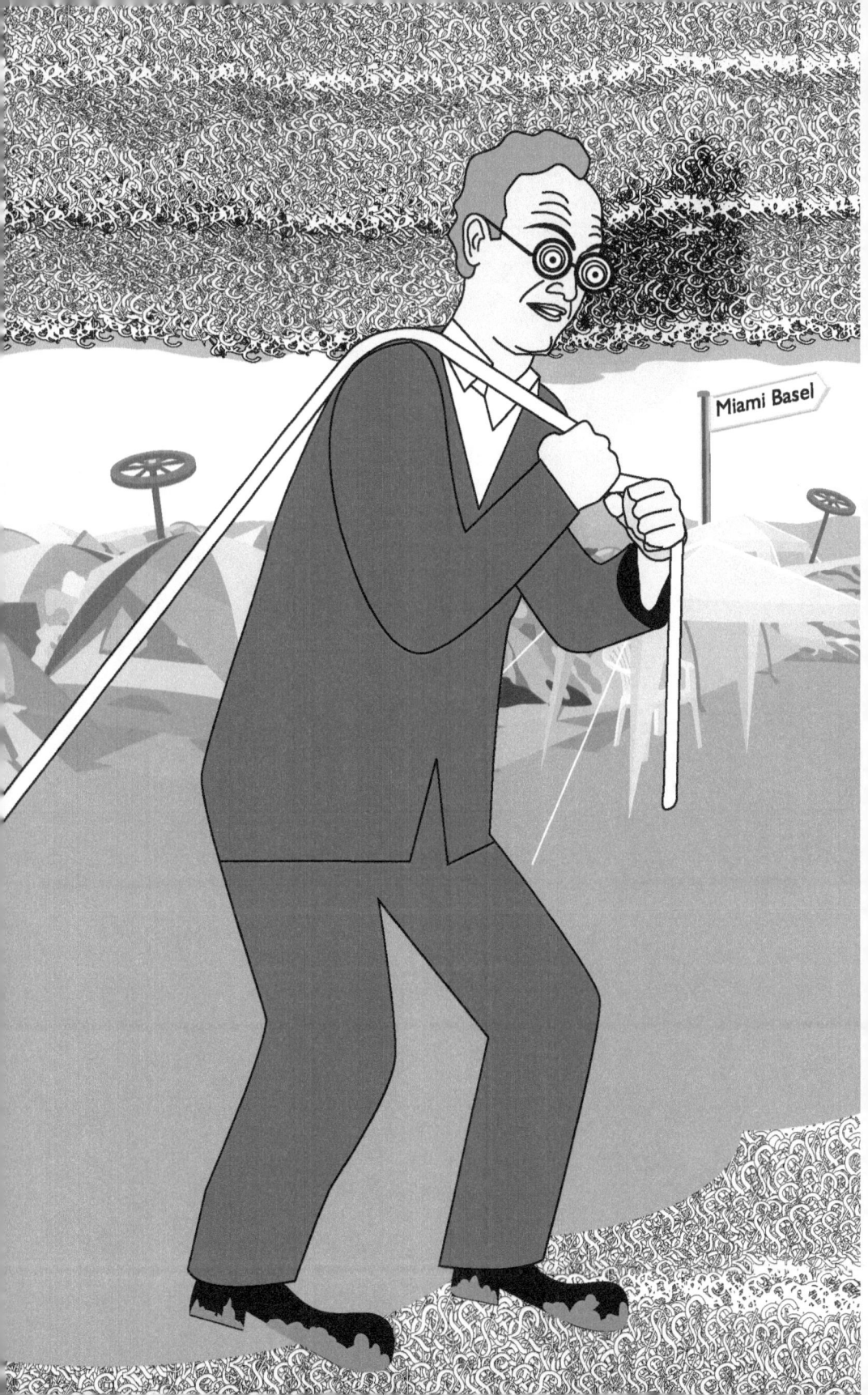

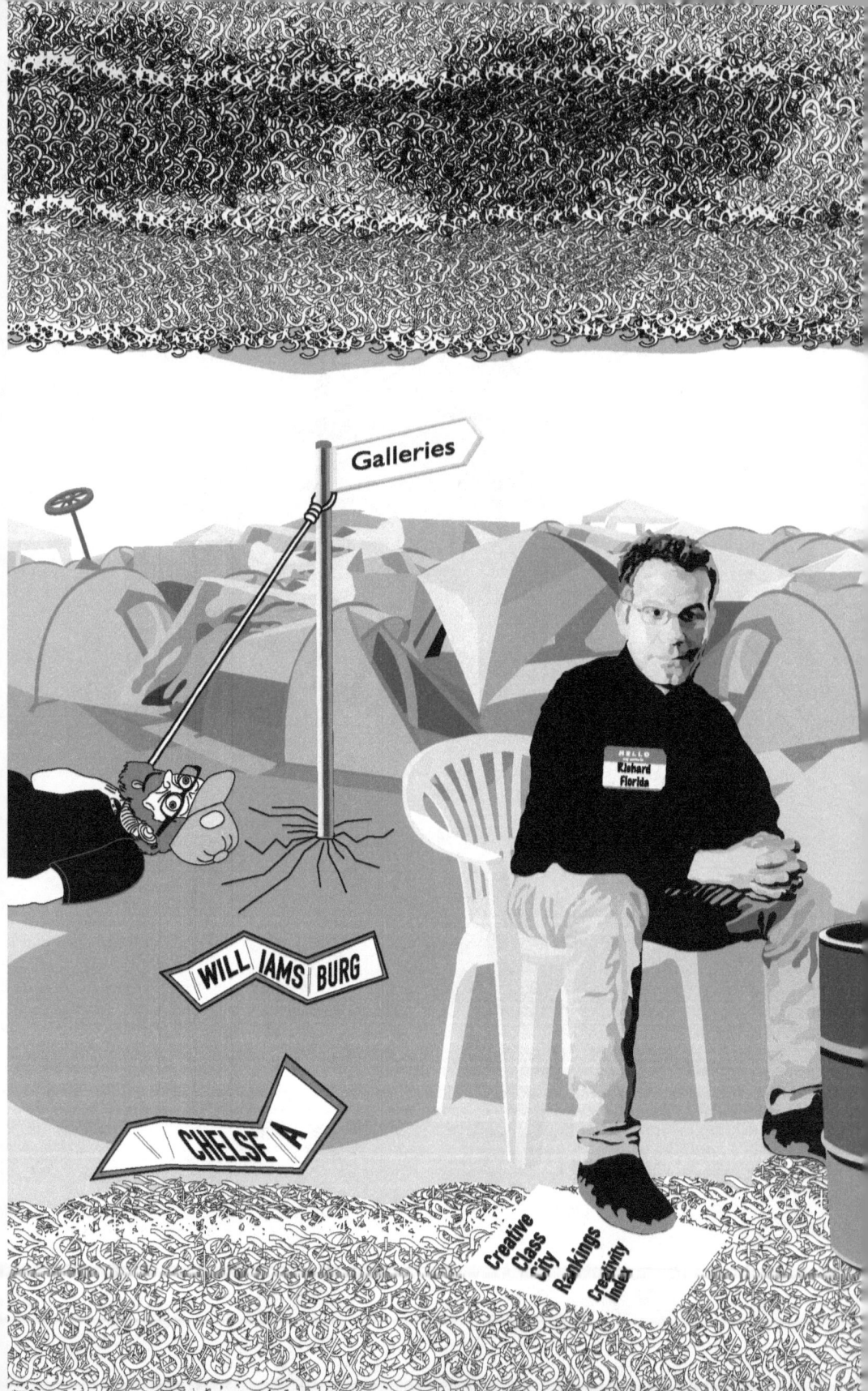

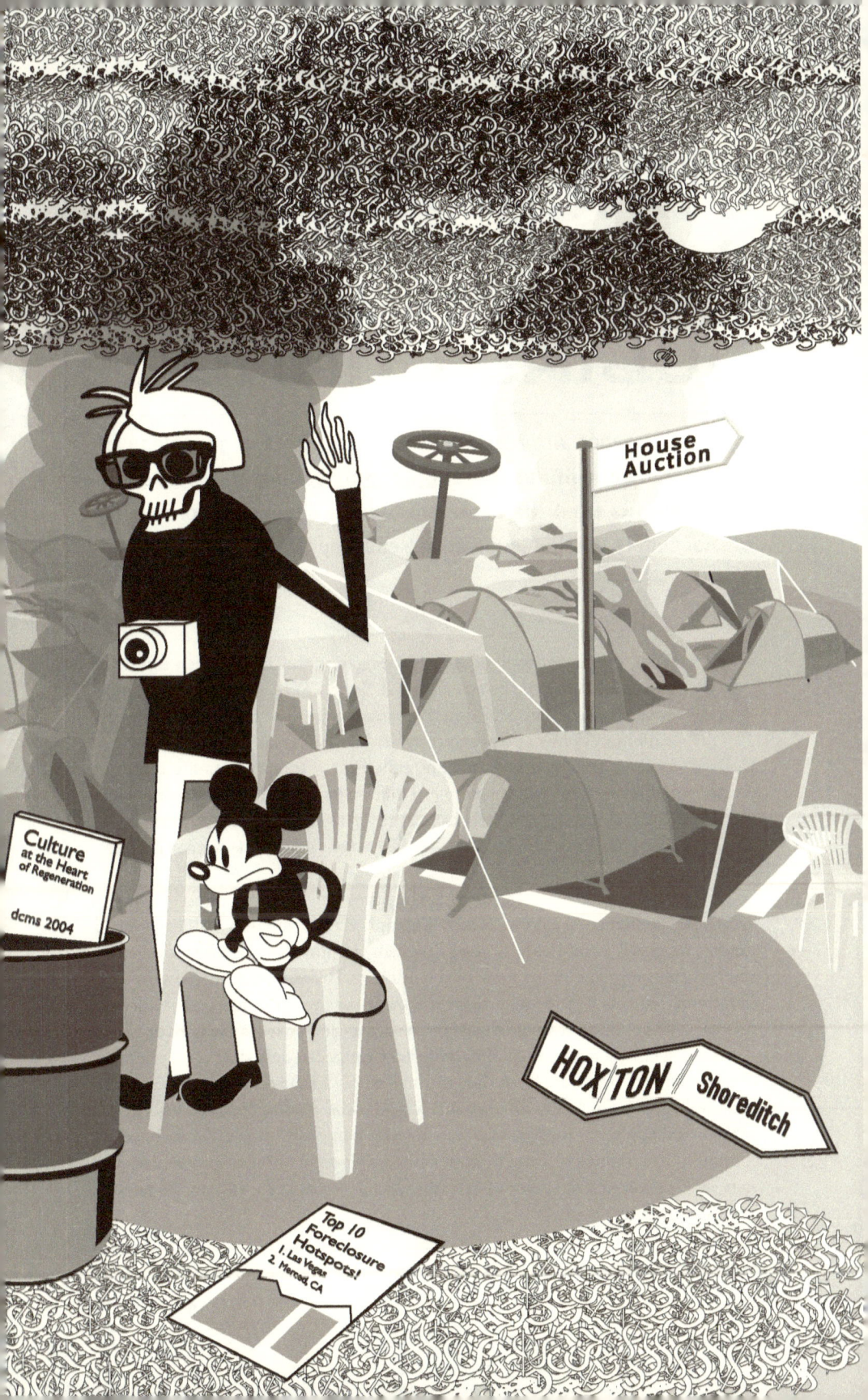

CONCERNING ART AND SOCIAL CHANGE

The 2007 reader *Art and Social Change* offers a genealogy of today's radical cultures. Here, Brian Holmes and Marco Deseriis glean insights from the book into today's dilemma of producing critical culture within recuperative 'semiocapitalism'

Among the groundswell of books investigating the link between aesthetics and politics, *Art and Social Change: A Critical Reader* is particularly ambitious. Published in 2007 as a companion volume to the historical survey exhibition Forms of Resistance at the Van Abbemuseum in Eindhoven, Holland, the book features a wide-ranging collection of texts and manifestos, divided into four sections corresponding to four major watersheds in contemporary social and political history: the Paris Commune of 1871, the Soviet Revolution of 1917, the social uprisings of 1968 and the 1989 revolutions in the former Eastern Bloc.

Editors Will Bradley and Charles Esche have completed the anthology by inviting six contemporary critics (Geeta Kapur, Lucy Lippard, John Milner, Gerald Raunig, Marina Vishmidt, and Tirdad Zolghadr) to provide both a historical context and an interpretation for some of the readings. However, the interpretative framework remains light enough that the core of the project resides in the selection of historical documents produced by the artists and activists themselves.

Thus, a critical appraisal of *Art and Social Change* can only start from matters of inclusion and exclusion. Even though Bradley and Esche do not make their criteria explicit, it is fair to say that the anthology has been compiled following a genealogical approach. Rather than searching for a mythical origin and its historic continuity, a genealogy sets out to 'maintain passing events in their proper dispersion', as Foucault explained after Nietzsche. The genealogist focuses on the numberless beginnings,

Image: Emory Douglas, illustration for *The Black Panther Paper*, 1972

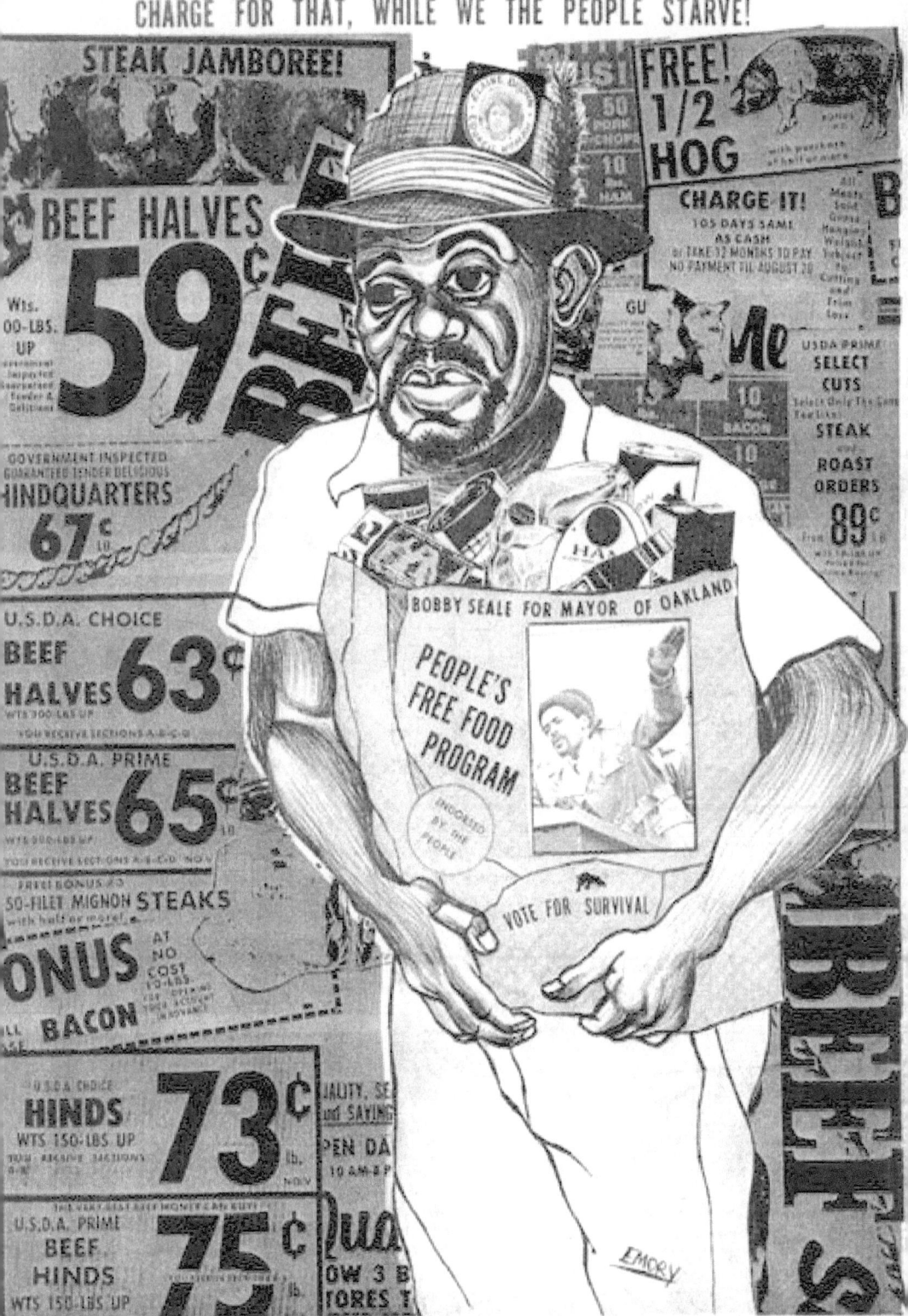

the accidents, the minute deviations – or conversely, the complete reversals, the errors, the false appraisals, and the faulty calculations that gave birth to those things that continue to exist and have value for us.[1]

By cutting through discontinuities and heterogeneous layers to isolate 'different points of emergence' in history, the genealogical approach avoids any identitarian closure, leaving the reader free to invent a trajectory through the material – and crucially, to decide upon its value in the present. However, Bradley's introduction does argue that the selection retraces geographically 'what might be termed the "globalisation of modernism"':

> The conception of art primarily at issue here is a modern, Western one that has been disseminated around the planet as the social, economic, and political conditions and institutions that support it have been replicated.[2]

This is only partly true, as the editors downplay or ignore the socially engaged art of Southern European countries such as Italy, Spain, and Greece, the decolonisation movements of the 1960s, and the entire Asian and Australian continents where the globalisation of modernism has been under way for several decades.

the post-'68 left refuses to deal with the full complexity of social relations

About a half of the selected authors are from the Americas, most notably the United States, and, in a much smaller measure, Brazil, Argentina, Chile, and Cuba. On the other side of the Atlantic, the book pays due homage to the European avant gardes with a specific focus on France, England, Germany, the Netherlands, Russia, Poland, and a handful of artists from Eastern and Northern Europe (Slovenia, Serbia, Hungary, Denmark, and Sweden). Closing gestures toward Asia and the Middle East serve primarily as a reminder of territories still unexplored.

But if we accept that this is mostly a 'modern, Western' selection, it is nevertheless worth highlighting some thematic blind spots. Striking by its omission, for example, is the powerful mural painting movement that began around Orozco, Rivera, and Siqueiros in Mexico and went on to inspire the socially engaged artists of the United States in the 1930s. This kind of gap is perfectly comprehensible if we read the anthology as a genealogy of post-'68 practices, turned decisively away from any mass address or modernising program, emphasising instead the autonomy and singularity of each experience. But if that is the case, then it would have been useful to include texts by the subcultures and creative fringes of social movements such as the Indiani Metropolitani of the Italian 1977, the activist side of punk, and all the experiments in guerrilla communication and culture jamming that stem from the No Future generation.

The scope of the book is broad enough to reveal surprising patterns and potential alliances among experiences that took place several decades apart and under quite different political circumstances. We shall now unpack a few of them by way of a conversation that will also include some of the missing references highlighted above.

Marco Deseriis: The book starts from a highly symbolic historic conjuncture, the Paris Commune of 1871. At the time, the bourgeois separation between art and social praxis was a *fait accompli*. The industrial revolution had just dealt a fatal blow to craftsmanship – an activity in which manual and intellectual skills, the pursuit of the aesthetic and the useful were still integrated. Once production was rationalised, artworks began to be identified only by their belonging to the aesthetic sphere. The theory of *l'art pour l'art* reflected this status quo: the removal of art from practical life and the tendency of artworks to lose their social function. With the Commune, however, the bourgeois autonomy of art came under scrutiny. Gustave Coubert's call to artists to take over museums and art collections in the course of the uprising, and William Morris' socialist conception of art 'as a necessity of human life which society has no right to withdraw from any one of the citizens', tell us that by the end of the 19th century artists had begun to reclaim a social function for art, in alliance with the workers' movement.[3]

Brian Holmes: The anthology reflects on its own departure point by including the Situationists' 'Theses on the Commune' from 1962. For them it was a gigantic festival marked by leaderless spontaneity, clearing the ground for a kind of 'unitary urbanism' by destroying the monuments of domination. Yet they also saw Courbet as a deluded idealist who toppled the Vendôme Column while ignoring the nearby Bank of France, ripe for looting. What really counted for the Situationists were the material results and the live aesthetic experiences of the Commune – an attitude passed on to the direct-action movements of 1990s.

There could have been another departure point, however: the 1848 revolutions, a fully-fledged cycle of struggles stretching all over Europe. In France, 1848 marked the first time urbanised workers recognised themselves as a class subject to unemployment, homelessness and starvation, and the first time they forced themselves onto the public stage of representative democracy. It's the archetypal contemporary conflict, experienced by factions of the entitled classes as a crisis of legitimacy that drove them into new political solidarities and utopian aesthetic

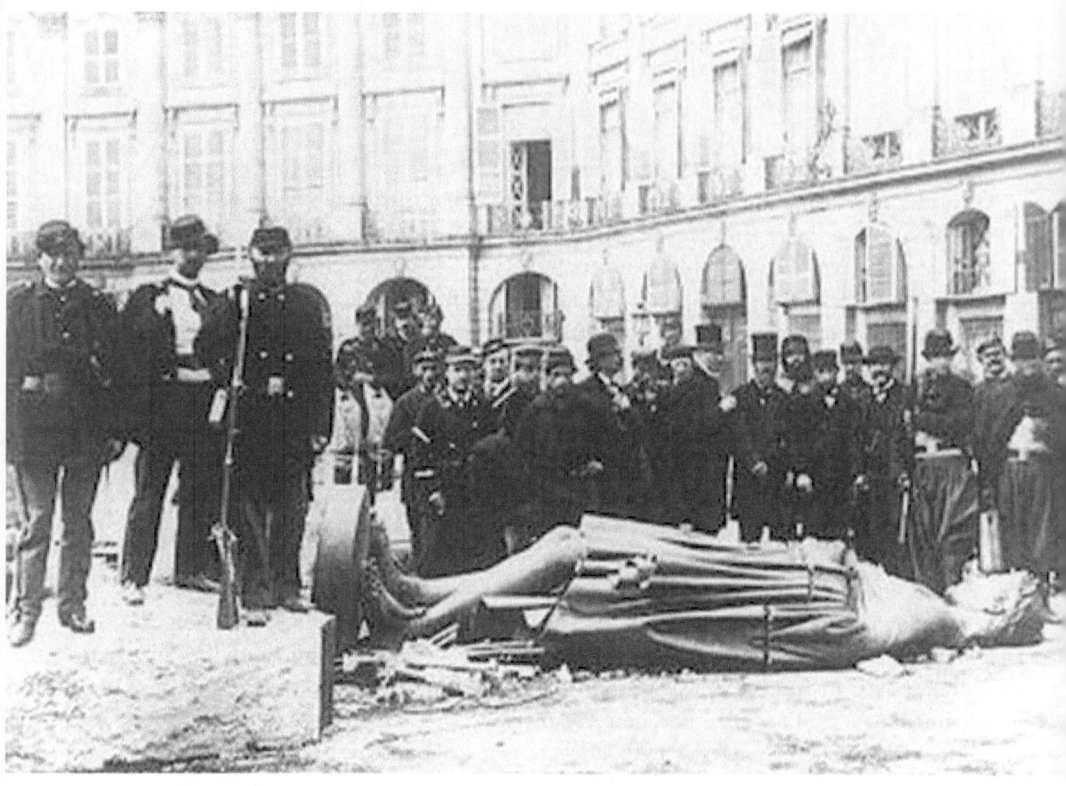
Image: Destruction of the Vendôme Column, Paris, 1871

experiments. It's clear why the editors of the anthology didn't want to sift through the flood of romantic idealism produced by that brief outburst of bourgeois experimentation. Yet here is a problem I see throughout the post-'68 left: a refusal to deal with the full complexity of social relations, and with the ambiguous or even disfigured forms they inevitably leave behind.

MD : Right, the social composition of the revolutionary wave of 1848 was very diverse and stratified. It included the urban petty bourgeoisie agitating for liberal reforms and national independence, as well as dispossessed farmers and factory workers. The thinker of the time who tried to integrate these social contradictions in a powerful aesthetic and political vision is Charles Fourier. Even if the phalanstères were designed to host the rich and the poor, and to restore social harmony without questioning the basis of capitalist accumulation, still they inspired many utopian communitarian experiments to come. Fourier conceived of labour as a pleasurable, playful activity to be modelled after human attitudes and desires. This made his theories quite appealing to the Surrealists and the Situationists, and in general to all social movements which place a high value on the productive power of imagination.

Brian Holmes and Marco Deseriis

Fourier's vision, which is unfortunately left out of this anthology, could be seen as an early instance of what Luc Boltanski and Eve Chiapello have called the 'artistic critique' of capitalism. Mostly developed by intellectuals, bohemians and exiles of the bourgeoisie, this critique targets the alienation and oppression derived from the Fordist organisation of labour and industrial specialisation of functions. It has been historically accompanied by the working class critique of the inequalities which stem from the unbounded accumulation of wealth characteristic of capitalism. But since this 'social critique' of inequalities does not necessarily imply a critique of alienation and oppression, and vice versa, Boltanski and Chiapello note that, depending on historical circumstances, these two types of critique may find themselves in association or in collision.[4]

> **by granting economic concessions and subsuming artistic critique, capitalism enters its late phase**

The social critique is represented in the anthology by two separate strands. The first runs from the Commune to Berlin Dada, the Surrealists, and after WWII, the Situationists International (SI) and the San Francisco Diggers. These groups all share the belief that art can be fully realised only by being abolished as an autonomous sphere – a task that for them is inextricably tied to the abolition of capitalism. The second strand is represented by the Bauhaus and De Stijl, which try to overcome the dichotomy between manual and intellectual work, functionalism and pleasure, by introducing new design principles into industrial production. In one respect, Russian constructivism can be associated with this second modernist strand in that it also tries to merge art and industrial production; yet on an ideological level it is closer to Dada and the Situationists in its fierce rejection of any style or aesthetics which is not subjected to the transient, historic needs of the working class.

BH: All that is well said, but to locate social critique in movements that sought either to turn art into everyday life or to fuse art and industrial design is to underscore what's been left out of this genealogy. Variations on socialist realism were spread around the world by the 1936 Popular Front against fascism, then reinvented in surprising ways by Third World liberation movements, often via surrealism. But they are barely represented here, so it's hard to grasp the affinities between the Black Panther graphics of Emory Douglas in the US, the murals of the Brigadas Ramona Parra in Chile, and the posters of the Atelier

Populaire in France. The missing link is revolutionary Cuba and the Tricontinental movement, which produced an influential series of posters advocating global revolution. The inclusion of the 1972 'Call to the Artists of Latin America' gives an inkling of the leftist politics, but not of its visual traditions. Well, the editors found the genealogies of direct action and self-organisation more compelling, and so do I. Still, there have been fantastic political experiments with the popular languages of street art, which bear much closer relation to the workers' movement and its social critique. Contemporary groups spring to mind, like Ne Pas Plier in France, or the Grupo de Arte Callejero and the Taller Popular de Serigrafía in Argentina, not to mention ongoing muralist movements in cities like Los Angeles.

MD: Yes, the murals and the political posters you mention (to which I would add the anarchist posters of the Spanish Civil War) are only marginally represented here. The reader seems to privilege expressive forms such as performances, manifestos, and calls to action. Of course, activist murals and posters can also incite action, but the visual representation of resistance is always at risk of being instrumentalised by political vanguards, or of being turned into folk art after its ties to a living social struggle have been severed.

In this respect, Debord's analysis of the spectacle may still be relevant. The reader features another situationist text, explaining that the expulsion of 28 members was due to their refusal to renounce their artistic careers – a position deemed incompatible with the SI.[5] This highlights a contradiction that keeps haunting activist art. If social movements have always expressed their imagining activity in one form or another, the culture industry puts these representations at risk of being commodified. As you noted in 'The Flexible Personality', Boltanski and Chiapello argue that it was by neutralising social critique through economic concessions and by subsuming artistic critique that capitalism was able to enter its late phase.[6]

BH: The anarcho-libertarian vein of the '68 movements was selectively mined for whatever could fit into the emerging hegemony of neoliberalism. This cultural integration is the glue that holds together a disjointed system. Boltanski and Chiapello were able to document the process by a statistical analysis of '68-era terms being used in managerial literature. But they see no value in artistic critique. They lament the decline of the workers' movement and they idealise the welfare period without questioning the welfare-warfare state, but they offer no resources for the characteristic struggles of the present.

By the late 1990s it was clear that whole new dimensions of alienation were setting in, not only on the receiving end of surveillance and monitoring, but at every level of professional and freelance practice. Individuals were constantly enjoined to be more productive, more appetitive, more intensively plugged into the digital prosumer system. All that was unimaginable without new developments in what I call the semiotic economy, referring both to the frenetic consumption of signs and images and to the continuous creation of credit money on the financial markets – two inseparable phenomena. Since the dead ends of this 'new economy' were already visible, I thought that artists should both critique the contemporary forms of alienation and find ways to revolt against them, in solidarity and collaboration with more directly oppressed and exploited people.

MD: The counter-culture was not only selectively mined to reorganise workflows around teamwork and flexibility, but also to harness its ethics of openness, sharing and decentralisation in the service of primitive accumulation in the IT sector. Fred Turner has shown how the encounter of San Francisco flower power with the technological culture of the Silicon Valley gave birth to what Richard Barbrook and Andy Cameron have labelled as the Californian Ideology, the 'anti-statist gospel of hi-tech libertarianism', which provided the 'spirit' of the late 1990s dotcom boom.[7]

In this respect *Art and Social Change* tells only the first part of the story. No doubt, the Diggers' 'post-competitive game' to build Free Cities across America, and their hilarious plan of action to persuade corporate employees to abandon the workplace – summarised in the slogan 'Give Up Jobs. Be With People. Defend Against Property' – offers a powerful blend of social and artistic critique.[8] But after purging this critique of its radical aspects, flexible capital offered workers the option of fleeing the office (but not work) through the new communication technologies; to 'be with people' by expressing their own creative and relational qualities in teamwork; and to overshadow the issue of *property* and distribution of wealth under the ideology of the gift economy and free access to the information society. The result is that nowadays everyone is entitled to have free internet access, no matter whether they are homeless or survive on food stamps!

The largely untold side of this story, or its 'obscene underside' as Žižek would call it, is that the creative class quickly gentrifies the new urban districts where internet access is free but housing prohibitively expensive. Over the last two decades the cultural production of Richard Florida's 'high bohemians' has been appropriated and re-sold at exorbitant prices by the real estate industry: in spite of their low income, artists are no longer seen as potential allies by the working class, but simply as the spearhead of gentrification.[9] I think we need to ask ourselves how to reverse this trend. Where would you start from?

BH: I started by using new media to take to the streets, to spark confrontations. The thing is, spectacle has acquired a different meaning since the advent of flexible accumulation. It is no longer the homogeneous mass-produced image that characterised the state and corporate media of the 1950s and '60s. Now it is something you constantly produce through a plethora of miniaturised devices. It has to be critiqued in motion, at the heart of interactivity.

This anthology contains great insights into the sources of what is now called 'tactical media', particularly in a series of texts running from Cildo Meireles' 'Insertions into Ideological Circuits' to Paul Ryan's 'Cybernetic Guerrilla Warfare'. The meeting between multimedia computers and the concept-art strategies for the activation of the viewer is directly prefigured in the alternative video movements envisioned by the journal *Radical Software* in 1970. Just as Meireles suggested in the Brazilian context, it is a matter of inserting subversive expressions into the very circuits where life under late capital is configured. To appropriate the media is tantamount to appropriating the means of production because, as Félix Guattari pointed out, what's at stake is the production of subjectivity.[10]

> **artists are no longer seen as allies by the working class, but as the spearhead of gentrification**

Later developments can be glimpsed in the glossary of net-culture terms by the Raqs Media Collective or the interview with Ricardo Dominguez, one of the founders of the Electronic Disturbance Theatre. His work resonates with the hybrid border-crossing art of Coco Fusco and Guillermo Gomez-Peña, and with the 'invisible theatre' of Augusto Boal in 1970s Brazil, which shifted Brecht's alienation-effect into everyday spaces of performance. Figures like Boal are sources for the performative politics of the Zapatistas, with whom Dominguez has closely cooperated. So even in the most high-tech, networked performance there is a transformation of an aesthetic and revolutionary history stretching back to the Third World movements, which themselves reworked the internationalist struggles against fascism. If the anthology had included some links between the 1970s in Italy and the counter-globalisation movements, then we could trace similar lines of development in and around Europe. Unlike Boltanski and Chiapello, I don't see how cultural issues can be disentangled from the politics of life and labour. Yet there has been a real decline in the capacity of artists to arouse outrage at both alienation and exploitation. The ideological force of neoliberal culture has been amazingly effective.

Brian Holmes and Marco Deseriis

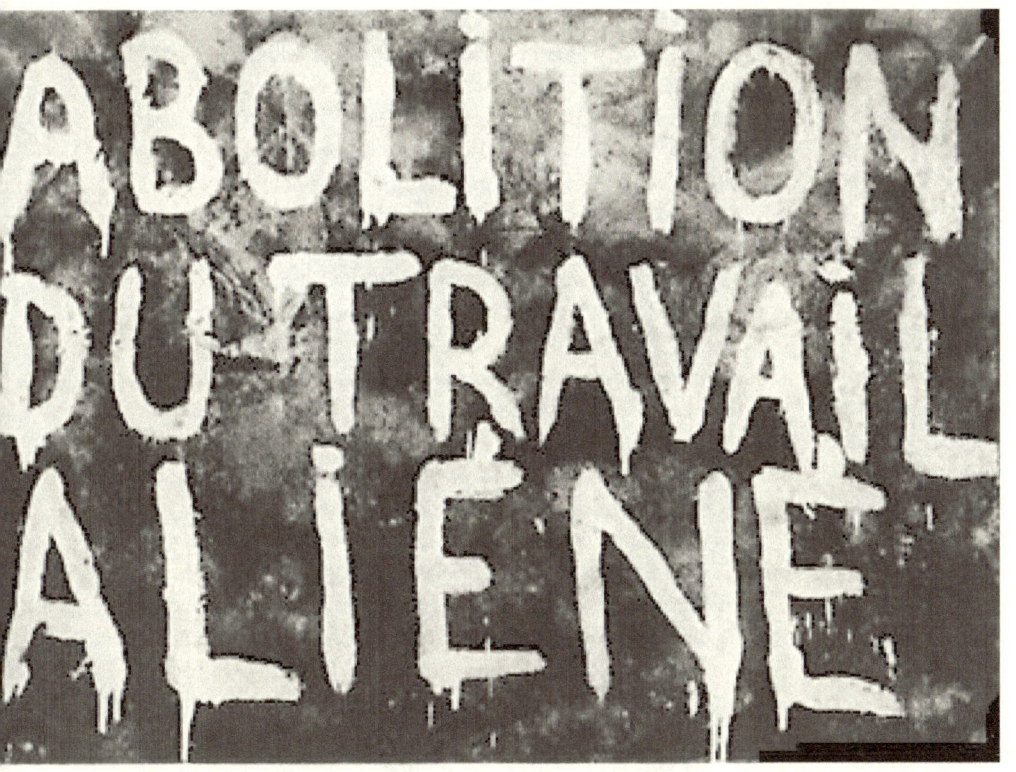

Image: Guy Debord and Giuseppe Pinot-Gallizio, *Abolition du Travail Aliéné*, 1963

MD : In my view, there are two major reasons why the critique of alienation has withered. The first one is that as the opportunities for activists to insert themselves into the 'ideological circuits' increased, the feeling of being just another spectator declined (Indymedia's motto 'Don't Hate the Media, Become the Media' perfectly exemplifies this shift). With the emergence of the network society it is increasingly problematic to claim, as Debord did, that the spectacle is simply the language of capitalist separation, that is, a one-to-many relationship which rigidly divides 'what is *possible* from what is *permitted*.'[11] In recent years, the single Spectacle has been accompanied by many spectacles which open up new political imaginaries while enhancing the autonomy of their producers.

The second reason why the 1960s artistic critique of alienation is no longer effective is that the new machines have become, as Donna Haraway once said, our 'intimate components' and 'friendly selves'.[12] Since these machines enable a strange mix of repetitive and creative tasks, manual and intellectual activities, revolting against them would be like revolting against ourselves. But if in the age of networked publics every social and linguistic activity is immediately put to work, how can we produce subjectivities that do not simply feed capital? The text 'How To?' by the Tiqqun group, which can be criticised for its ambiguous celebration of political anonymity and insurrectionalist mystique, advances the idea of the Human Strike.[13] From my understanding, this is a strike against the production of fake subjectivities and critiques which are not rooted in local communities sharing strong affective bonds. This could be read as a critique of the tactical media scene, and more broadly, of the tendency to generate critical discourses detached from any need to transform society from the ground up.

B H: 'How To?' is a poetic call to insurrection, resonating with a number of recent revolts. Tiqqun has developed a radical critique of cybernetic society, which they see as a total mobilisation of the ego for meaningless labour and consumption. Of course that implies a critique of tactical media, which has been largely neutralised by corporate and government sponsorship. Tiqqun proposes an anti-aesthetic, a way of cutting the feedback loops that bind us to our marketised selves. It is linked with a rising desire to exit the cities; but with an obvious difference from 1960s utopias, of the kind represented in the reader by the fascinating, yet little-known artist Bonnie Sherk with her text on the social art work of 'The Farm' in San Francisco. A more or less violent aesthetics of withdrawal from neoliberal networks is now being expressed in many countries, because it's one way to begin shifting towards a new paradigm. A very dangerous way, however – which could also justify further escalations of the military-police state and the durable installation of authoritarian neoliberalism.

In the early part of this decade, I hoped that networked social movements could theatricalise the real social relations of neoliberal society, suspending its functional norms and opening

> an artistic critique of alienation cannot ignore the fact that art is a form of power

Brian Holmes and Marco Deseriis

up space and time for collective questioning. That was done by unleashing an unauthorised circulation of subversive ideas, to be embodied by militant groups in urban situations. Or anyway, that's how I explained it in 'The Revenge of the Concept', included in this anthology. Those protests worked to a degree, because there has been an intense examination of neoliberalism including both social and ecological critiques. Now I'm hoping that social movements can influence the new economic paradigm that will ultimately emerge from the crisis, by inventing critical and poetic discourses, shaping viable territories and constructing cooperative machines. Leafing through the reader, I find myself wondering if our times can produce anyone like Theo Van Doesburg, who was both a dadaist and a member of De Stijl, and who called destructively for the end of art while designing a constructivist aesthetics of built environments. But that reference is a bit too genealogical! What do you see as the discontinuous breakthroughs of the present?

MD: I think we need to start by relocating the forms of alienation. As Matteo Pasquinelli points out in his new book *Animal Spirits*, the libidinal economy of the network society generates a symbolic surplus-value that the real estate market tries to parasite and put to work in the new creative districts of the world cities.[14] The fact that information is an abstract, non-rivalrous good does not mean that its wealth cannot be appropriated and congealed into (private) space. From this flows the notion that the critique of intellectual property cannot be separated from a critique of material property and of rent, that is, of the earliest form of capitalist property.

In the contemporary factories of biopolitical production, artists are the urban pioneers who identify the hotspots in advance. Therefore it is their duty to ask themselves how to retain the value they produce through their common imagining activities and cultural production *within* the community in which they live. The reader features a great article on the group Park Fiction, which describes the successful struggle of a Hamburg association of residents to design and obtain a public park through a participatory planning procedure in the historic district of St. Pauli, against the attempts of the city to gentrify it.[15]

On a symbolic level the critique of alienation can take on many different forms. Personally I am fond of collective experiments such as Laibach, Luther Blissett, etoy, Ubermorgen, Rtmark, Yo Mango! and others which subtract themselves from the hegemonic field by changing the very coordinates in which they operate. The music and performances of Laibach exemplify this strategy perfectly. By juxtaposing industrial sounds, pop songs, totalitarian symbols, and avant garde art, Laibach restages the foundational violence of the nation state and of every spectacle. As they say in

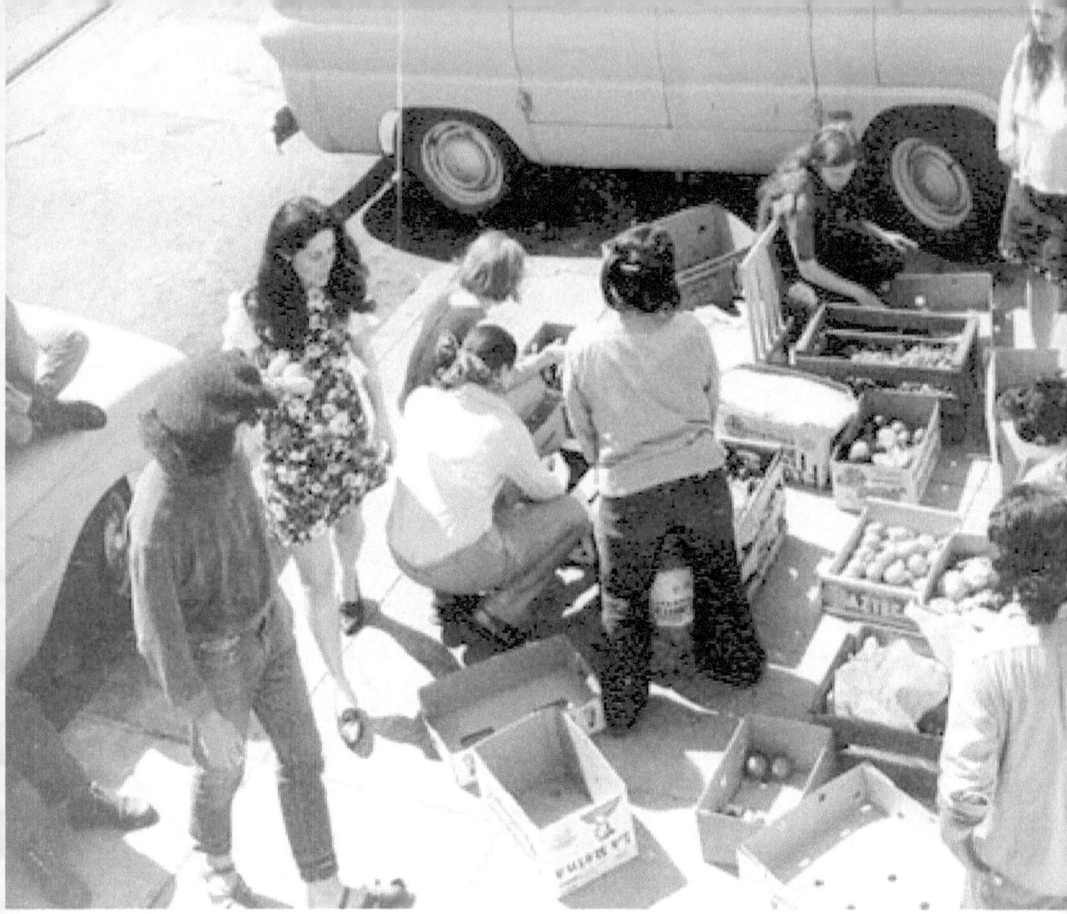

Image: San Francisco Diggers distributing free food, 1966

their manifesto, the *Ten Items of the Covenant*, included in the reader: 'All art is subject to political manipulation [...] except for that which speaks the language of this same manipulation.'[16]

It is important to recognise, however, that Laibach's performances do not only target pop culture and totalitarian ideology but also art itself, in particular the will to power of the avant garde. I believe that a new artistic critique of alienation cannot ignore the basic fact that art *is* a form of power. This is particularly true now that art is no longer used, as in totalitarianism, to aestheticise politics, but to redesign the urban space according to the shifting needs of semiocapitalism. In this respect it would be very important to update Lefebvre's analysis of the accumulation and distribution of power in urban space, to stimulate practices of reappropriation in which artists could play an important role. After all Lefebvre's *The Right to The City*, written in 1967, became the slogan of a whole movement.[17]

Info

Will Bradley and Charles Esche (eds.), *Art and Social Change: A Critical Reader*, Tate Publishing UK, 2008

Brian Holmes and Marco Deseriis

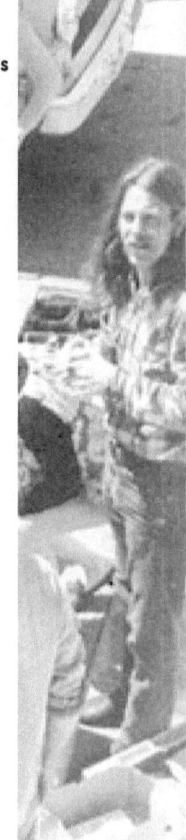

Footnotes

1 Michel Foucault, *Language, Counter Memory, Practice: Selected Essays and Interviews by Michel Foucault*, trans. and ed. Donald F. Bouchard, Ithaca, NY: Cornell University Press, 1977, p.146.

2 Will Bradley, 'Introduction', in Will Bradley and Charles Esche (eds.), *Art and Social Change: A Critical Reader*, London: Tate-Afterall, 2007, p.11.

3 William Morris, 'The Socialist Ideal: Art', in *Art and Social Change*, p.50.

4 Luc Boltanski and Eve Chiapello, *The New Spirit of Capitalism*, trans. Gregory Elliott, London: Verso, 2007.

5 Situationist International (J. V. Martin, Jan Strijbosch, Raoul Vaneigem, René Viénet), 'Response to a Questionnaire from the Center for Socio-Experimental Art', in *Art and Social Change*, pp.125-9.

6 Brian Holmes, 'The Flexible Personality: For A New Cultural Critique', in *Hieroglyphs of the Future*, Zagreb: WHW/Arkzin, 2002. Available at http://transform.eipcp.net/transversal/1106/holmes/en

7 Fred Turner, *From Counterculture to Cyberculture: Stewart Brand, the Whole Earth Network and the Rise of Digital Utopianism*, The University of Chicago Press, 2006; Richard Barbrook and Andy Cameron, 'The Californian Ideology', available on multiple websites, including www.alamut.com/subj/ideologies/pessimism/califIdeo_I.html

8 The San Francisco Diggers, 'Trip Without A Ticket', and, 'The Post-Competitive, Comparative Game of a Free City', in *Art and Social Change*, pp.146-56.

9 Richard Florida, *The Rise of the Creative Class: And How It's Transforming Work, Leisure and Everyday Life*, New York: Basic Books, 2002.

10 Félix Guattari, 'On the Production of Subjectivity', in *Chaosmosis: An Ethico-Aesthetic Paradigm*, Indiana University Press, 1995.

11 Guy Debord, *Society of the Spectacle*, trans. Ken Knabb, London: Rebel Press, 2002, p.14.

12 Donna J. Haraway, *Simians, Cyborgs, and Women: The Reinvention of Nature*, New York: Routledge, 1991, p.178.

13 Tiqqun, 'How To?', in *Art and Social Change*, pp.297-312.

14 Matteo Pasquinelli, *Animal Spirits: A Bestiary of the Commons*, Rotterdam: Nai Publishers, 2009.

15 Christoph Schäfer and Cathy Skene with the Hafenrandverein, 'Rebellion on Level P', in *Art and Social Change*, pp.283-9.

16 Laibach, 'Ten Items of the Covenant', *Art and Social Change*, p.251.

17 Henri Lefebvre, 'Right to the City', in *Writings on Cities*, trans. and ed. Eleonore Kofman and Elizabeth Lebas, Cambridge, MA: Blackwell, 1996, pp.63-177.

Marco Deseriis <snafu AT thething.it> is a PhD Candidate at New York University. He has co-authored a book on Net.Art (Shake, 2003-08) and collaborates with the festival of Culture Jamming and Radical Entertainment, The Influencers, http://theinfluencers.org

Brian Holmes <brian.holmes AT wanadoo.fr> is a cultural critic, living in Paris and Chicago. He is the author of *Hieroglyphs of the Future: Art and Politics in a Networked Era* (Zagreb: WHW, 2002) and *Unleashing the Collective Phantoms: Essays in Reverse Imagineering* (New York: Autonomedia, 2008). He currently collaborates on the Continental Drift seminar with the 16 Beaver group in New York. Forthcoming book and text archive at http://brianholmes.wordpress.com

ALL MOUTH, NO HISTORY

A purely linguistic analysis of financialisation and contemporary production fails to account for the current crisis, argues William Dixon in his review of Christian Marazzi's latest book

Perhaps the oddest thing about Christian Marazzi's book *Capital and Language: From the New Economy to the War Economy*, is that it has been published at all. On page 145 we find out that, 'As I'm writing this, [it is] exactly six months after the 11 September terrorist attack'. The book was indeed first published in Italian in 2002, but now this translated version appears in the shadow of a crisis that demands its own analysis. The question really is whether this book may help orientate us to the current episode. Of course, this is a somewhat unfair question given the scale of what we are confronting today. One focus of this book is on securities, and we shall return to that, but the current crisis is peculiarly about banking, securitised mortgages, housing, and government responses. There are relevant points that can be carried through, but just how far out of its depth, however involuntarily, this book is, may be seen in its discussion of the strategy to reduce government debt and to financialise pensions. The current crisis can be judged by the dramatic turnaround on government spending, on government debt as a percentage of GDP, and on state control of aspects of the economy. Marazzi could, reasonably, respond that he has dealt with the turn to war following 9/11 as a crucial element of government spending, but really we have now entered a new phase. The prospect we face at the moment is the transformation of a private debt crisis (banks, mortgages and credit cards) into a government debt crisis as governments attempt to avoid getting locked into deflation. These outcomes aren't discussed in this book.

The thesis of the book is that language and communication are crucial both to production and to finance and that, 'it is for *this very reason* that changes in the world of work and modifications in the financial markets must be seen as two sides of the same coin' (p.14). The point is that work is developing in such a way that the general intellect resides with workers themselves and is not simply encrusted in fixed capital. In finance, especially as we look at shares, their movement depends on what others are doing; there is a reflexivity here that means conventions are crucial, as are language and communication.

On the finance point, we are looking at territory already marked out very distinctly by Keynes and that has been explored in different ways by people such as Minsky or, more recently, Schiller, who identifies a financial contagion as the cause of the crisis. Marazzi quotes Keynes and Schiller but appears, with his focus on language, to be

William Dixon

adding something to these writers. It is not clear, however, whether anything meaningful is added. Keynes argued that in financial markets, knowing conventions may be more important than fundamental knowledge of the matter in hand. So communication, language, is indeed important. The question in the study of financial matters is whether conventions can really overcome fundamentals, i.e. issues such as profitability. This question is clouded, though, since it is really future profitability that is at issue and that, of course, can be subject to all manner of thoughts and speculations that will be prone to conventions. The speculative dotcom bubble, in the aftermath of which Marazzi was writing, was built on a developing convention boosted by all manner of communications. Of course this did indeed fall very flat. Once the price earning ratio reaches absurdly high levels, as it did during the dotcom bubble, the bet on future earnings is so extraordinary that fundamentals must reassert themselves. The eventual fall put the so called 'real economy' at risk because the investment of pensions in shares resulted in a need for individuals to boost their pension pots with extra savings, hence withdrawing them from consumption.

Given this potential for seemingly aberrant fluctuations of markets, how can we set out an understanding? For Marazzi, 'to explain the workings of financial markets in the era of post-Fordism what we need is a linguistic theory of their operations' (p.29). What follows is a study of language analysis 'from the point of view of its *biological foundation*' (p.29). Discussion of language is interesting, but what use is this to understanding financial markets? One point Marazzi makes is that language may be performative, in the sense of performing rather than informing. Basic performativity is seen in the statement, 'With this ring I thee wed'. It produces the act. Marazzi then refers to the absolute performativity of the statement, 'I speak', in which what is produced is purely a language event. This is an 'especially useful' category for Marazzi because 'it is immediately applicable to the crisis of the financial markets as a crisis of the *overproduction of self-referentiality*' (p.35).

This does point to some reality in financial markets. However to rely on a philosophy of language at this point tends to mystify what has happened. Marazzi writes just at the end of the dotcom boom. Like previous financial manias there was a self-referential element, the contagion, in that share prices rose partly on the then objective basis that share prices were rising. The bubble is based on a convention that has an objective basis in the sense that it is autonomous from any particular individual, even resisting the attempts by Greenspan to talk it down. The theoretical understanding of this phenomenon was already outlined in Keynes' *General Theory*. Keynes' explanation had the considerable virtue of being better rooted in historical developments. The starting point was the separation of ownership and control through the introduction of joint stock companies and limited liability that in turn allowed the introduction of a proper market in company shares. A company could raise capital through share offers setting up three distinct groups: the shareholders, the managers and the entrepreneurs. The share made entry to investment easier because subsequent exit (sale) was easier and so allowed the raising of capital for large projects. This securitisation meant that individuals (but not societies) could revise decisions on investment ownership without entailing the sale of physical assets. This enabled the development of a portfolio capitalism in which risk was no longer particularised but could be subdivided and then diversified. Individual capital could aspire to be free from engagement through particular capital with particular labour power. We might say that this was in line with the tendency of finance, developed considerably further in recent years, to aim to be free to accumulate without being tied down.

In these terms, then, the nightmare is that capital must be validated, in its circuit, through the purchase of labour power. At this point it appears to be at risk. Securitisation seems to offer the compromise solution. The combination of limited liability and shares allows diversification that could free capital from one particular labour force, or indeed even one national labour force. This is where the self-referential element of financialisation enters the picture; the market in shares, that separation, in principle, of ownership and control and so also freedom from specific knowledge based responsibility, posed a problem of valuation. Keynes' argument was that the nature of investment in bourgeois society was inevitably uncertain because, with the separation of production from consumption, the future was unknown. The entrepreneur overcame this through gut instincts or 'animal spirits' that were founded on some degree of domain knowledge involving the labour force, the consumers, etc. Such an entrepreneur could value future prospects on the basis of real, even if not guaranteed, knowledge. Securitisation undermined this because in opening a market, the value of the security depends on how others view the current prospects. Because of the inherent uncertainty along with the possibility of flight, i.e. sale, and because investors may not have domain knowledge, they are subject to 'news' which can provoke rapid revaluations of shares. This puts the entrepreneur in a curious position when considering decisions about shares. S/he may consider the worth of a share on the basis of informed gut instincts, but will know that the actual price of a share will vary according to the estimations of everyone else, news, etc. This sets up what Keynes referred to as a 'beauty contest'. I may know who I think is most beautiful but to work out who will actually win I have to work out who other people think will win. Now, consider that when investors make their judgements, they do so on the basis of what they think other investors will think, and vice versa. This can, as Marazzi suggests, be described as a problem of self-referentiality but it doesn't require a discourse on language to get there.

There is a problem of anticipating what other people will do when they are anticipating what I will do; this may result in Hobbes' problem – a war of all against

William Dixon

all and, hence, the absence of community. Evidently, financial markets cannot work like that and nor can they resort to Hobbes' solution – the absolute sovereign – since this only recasts the problem at another level. It is a real insight to say, as Marazzi does, along with Keynes, that financial markets require conventions. It is also right, as Marazzi says, that this convention forms the community (and vice versa) which in this case is the market itself. The convention here concerns the valuation of securities, which in turn concerns the prospects for the future that can seem relatively stable but can also change with appropriate news that provokes dramatic revaluation. Whatever any one individual may think about the news, the key is to understand how others understand it since valuation depends on convention. In these circumstances news can take on an exaggerated importance. This has become a matter of great importance recently. We know that an aspect of the sub-prime market was the selling on of mortgages that were then, in turn, engineered into financial objects which were packaged up and sliced up to create mortgage backed assets. They were held by a number of institutions including, of course, banks. So far so good, in a manner of speaking. Once people began to renege on those mortgages and the complication of the financial objects came to the fore, then the exposure of risk became unclear. At a certain point no convention could hold sway with those securitised assets and no market could be made. It doesn't matter at this point that certain banks may still think there is value in these assets.

The failure to make a market may be seen as the collapse of a certain kind of community. This represents a serious crisis for bourgeois society. Banks faced a crisis in terms of balancing between assets and liabilities, so not only did banks want to hold on to cash but they also didn't want to lend to other banks that were essentially in the same peril as themselves – so the credit crunch. This had, of course, already induced an anticipatory collapse of shares as everyone began the quest for cash. Here was the irony of the situation. Securitisation, by which individual agents diversified risk, had locked everyone into the crisis. Securitisation enabled portfolios, but no portfolio could allow for the level of systemic risk. You can't diversify out of the system except through cash, but that then is part of the problem. The portfolio, in socialising returns, simply moved the possibility of crisis to a wider level. The result now with credit contracting is a deflation that could well become long term stagnation. Ironically also, with the liquidity subsequently pumped into the system, with banks holding dubious assets, and with governments now taking on large debts, we have the preconditions in terms of the supply side (money) and even demand for inflation (solving the problem of toxic assets and government debt by turning the telescope around) for a quite dramatic inflationary episode, if there is some move to recovery.

The point about the above is that it gives an account of self-referentiality with a historical basis. It used to be a strength of autonomist analysis that it could ground theoretical developments in historical developments. It is no leap forward to replace, as Marazzi does, history with the philosophy of language. Keynes, in maintaining some historical perspective, gets further. The analysis can be developed further still. Securitisation can be related directly to the nature of class dispositions. The portfolio allows not just the development of machinery but also some outflanking of particular working class demands. These two points are essentially the same, though, since the rising organic composition, machinery, is anyway a response to the 'refractory hand of labour', and just as this poses the issue of raising a larger scale of capital, it also poses the need to diversify risk.

The use of machinery also relates to that other aspect of Marazzi's story – the rise of the knowledge worker for whom language is important. The presentation of this knowledge worker would also have been considerably deepened with a historical perspective. In what way is this worker different from the skilled workers of the past, of the start of the 20th century? They were replaced, restructured, through scientific management and assembly lines, or what Marazzi refers to as Fordism. A study of F.W. Taylor and then H. Ford shows that the reformulation of work practices from the late 19th to early 20th centuries was intended to shift control from workers to management, and a key element of this was the introduction of explicit measurability. This led to the massification of the worker and the subsequent turn against work. Marazzi suggests something entirely new has occurred today, since the general intellect is located with knowledge workers, posing the problem of control and measurability. It would help if there was an explicit comparison with the previous period of the early 20th century and those skilled workers by whom a project of workers' control had been imagined. A comparison would help us orientate ourselves. The same applies to Marazzi's discussion of the expansion of information under the 'New Economy' that is attended also by an expansion of work time, hence the problem of an attention deficit. All of this needs quite a bit more working out, since it is not clear whether this problem, however interesting, is new or even important. Like much of the rest of the book, there is here the suggestion of importance through a rhetorical strategy of radical nominalism, a dependence on naming that boosts the credentials of the work but not necessarily the analysis. We are given some familiars like Fordism, and post-Fordism to which are added the New Economy, Empire (Negri and Hardt), the attention economy and the war economy. These are backed by a host of supporting ideas, such as 'the overproduction of self-referentiality', that are signalled to us by the imperious use of *italics*. This is a path by which the rhetorical strategy separates profundity from understanding. I would have preferred more analytical explanation and less naming.

Of course, we must always attempt some synthetic view when faced by a separation of disciplinary approaches that appear to exclude what matters – namely, us.

I'd suggest that history provides a surer basis. The work of Sergio Bologna on class composition, for example, is solidly based. Even Negri's earlier work on Keynes is a start in this regard. A combination of history and analysis can keep our feet on the ground. We can take as an example the issue of pensions, which is undoubtedly a significant issue. Marazzi has this to say:

> The diversion of savings to securities markets, initiated by the 'silent revolution' in pension funds, has just this objective: to eliminate the separation between capital and labour implicit in the Fordist salary relationship by strictly tying workers' savings to processes of capitalist transformation/restructuring. (p.37)

There are a number of problems here. Pension funds, even if they tie workers in the way suggested, also tie capitalist performance to consumption. So if the objective is to eliminate the separation of capital and labour, how it is achieved remains ambiguous. Also how would this elimination of separation be any different from the salary relationship? Workers were as tied to capitalist restructuring by the wage as by any pension. Indeed pensions can be argued to have delivered considerably less than was promised. The more interesting thing in the USA (and the UK) in particular is the shift of emphasis from the wage to consumer debt. Cheap money and liquidity have been only part of the story here. More important was the deregulation of credit. This was a form of privatised Keynesianism. It allowed rising living standards, a developing financial sector, rising imports – ultimately from China – and also, of course, the great dream of home ownership, considered to be the crucial stake in society. Here are all the ingredients of the current reckoning. In the UK, the Tories are distancing themselves from all of this, despite having started it, by blaming Brown and so giving themselves the ideological space to impose severe austerity should they get elected next time. Marazzi refers to a liquidity crunch in October 2000 that 'until then had seemed unimaginable' (p.69). That crunch was dealt with with relative ease, but the current one really is 'unimaginable' and needs something more than a philosophy of language to get to grips with often mind boggling figures and developments.

Info

Christian Marazzi, *Capital and Language: From the New Economy to the War Economy*, translated by Gregory Conti, introduction by Michael Hardt, New York: Semiotext(e), 2008

> William Dixon <w.dixon@londonmet.ac.uk> researches the nature of economics, especially in relation to morality, political considerations and the development of society

DEBT: THE FIRST FIVE THOUSAND YEARS

Anthropologist David Graeber argues that it is only with a general historical understanding of debt and its relationship to violence that we can begin to appreciate our emerging epoch. Here he begins to fill in our historical knowledge gap

What follows is a fragment of a much larger project of research on debt and debt money in human history. The first and overwhelming conclusion of this project is that in studying economic history, we tend to systematically ignore the role of violence, the absolutely central role of war and slavery in creating and shaping the basic institutions of what we now call 'the economy'. What's more, origins matter. The violence may be invisible, but it remains inscribed in the very logic of our economic common sense, in the apparently self-evident nature of institutions that simply would never and could never exist outside of the monopoly of violence – but also, the systematic threat of violence – maintained by the contemporary state.

Let me start with the institution of slavery, whose role, I think, is key. In most times and places, slavery is seen as a consequence of war. Sometimes most slaves actually are war captives, sometimes they are not, but almost invari-

Images by Nicholas Brooks

David Graeber

ably, war is seen as the foundation and justification of the institution. If you surrender in war, what you surrender is your life; your conqueror has the right to kill you, and often will. If he chooses not to, you literally owe your life to him; a debt conceived as absolute, infinite, irredeemable. He can in principle extract anything he wants, and all debts – obligations – you may owe to others (your friends, family, former political allegiances), or that others owe you, are seen as being absolutely negated. Your debt to your owner is all that now exists.

This sort of logic has at least two very interesting consequences, though they might be said to pull in rather contrary directions. First of all, as we all know, it is another typical – perhaps defining – feature of slavery that slaves can be bought or sold. In this case, absolute debt becomes (in another context, that of the market) no longer absolute. In fact, it can be precisely quantified. There is good reason to believe that it was just this operation that made it possible to create something like our contemporary form of money to begin with, since what anthropologists used to refer to as 'primitive money', the kind that one finds in stateless societies (Solomon Island feather money, Iroquois wampum), was mostly used to arrange marriages, resolve blood feuds, and fiddle with other sorts of relations between people, rather than to buy and sell commodities. For instance, if slavery is debt, then debt can lead to slavery. A Babylonian peasant might have paid a handy sum in silver to his wife's parents to officialise the marriage, but he in no sense owned her. He certainly couldn't buy or sell the mother of his children. But all that would change if he took out a loan. Were he to default, his creditors could first remove his sheep and furniture, then his house, fields and orchards, and finally take his wife, children, and even himself as debt peons until the matter was settled (which, as his resources vanished, of course became increasingly difficult to do). Debt was the hinge that made it possible to imagine money in anything like the modern sense, and therefore, also, to produce what we like to call the market: an arena where anything can be bought and sold, because all objects are (like slaves) disembedded from their former social relations and exist only in relation to money.

But at the same time the logic of debt as conquest can, as I mentioned, pull another way. Kings, throughout history, tend to be profoundly ambivalent towards allowing the logic of debt to get completely out of hand. This is not because they are hostile to markets. On the contrary, they normally encourage them, for the simple reason that governments find it inconvenient to levy everything they need (silks, chariot wheels, flamingo tongues, lapis lazuli) directly from their subject population; it's much easier to encourage markets and then buy them. Early markets often followed armies or royal entourages, or formed near palaces or at the fringes of military posts. This actually helps explain the rather puzzling behavior on the part of royal courts: after all, since kings usually controlled the gold and silver mines, what exactly was the point of stamping bits of the stuff with your face on it, dumping it on the civilian population, and then demanding they give it back to you again as taxes? It only makes sense if levying taxes was really a way to force everyone to acquire coins, so as to facilitate the rise of markets, since markets were convenient to have around. However, for our present purposes, the critical question is: how were these taxes justified? Why did subjects owe them, what debt were they discharging when they were paid? Here we return again to right of conquest. (Actually, in the ancient world, free citizens – whether in Mesopotamia, Greece, or Rome – often did not have to pay direct taxes for this very reason, but obviously I'm simplifying here.) If kings claimed to hold the power of life and death over their subjects by right of conquest, then their subjects' debts were, also, ultimately infinite; and also, at least in that context, their relations to one another, what they owed to one another, was unimportant. All that really existed was their relation to the king. This in turn explains why kings and emperors invariably tried to regulate the powers that masters had over slaves, and creditors over debtors. At the very least they would always insist, if they had the power, that those prisoners who had already had their lives spared could no longer be killed by their masters. In fact, only rulers could have arbitrary power over life and death. One's ultimate debt was to the state; it was the only one that was truly unlimited, that could make absolute, cosmic, claims.

if slavery is debt, then debt can lead to slavery

The reason I stress this is because this logic is still with us. When we speak of a 'society' (French society, Jamaican society) we are really speaking of people organised by a single nation state. That is the tacit model, anyway. 'Societies' are really states, the logic of states is that of conquest, the logic of conquest is ultimately identical to that of slavery. True, in the hands of state apologists, this becomes transformed into a notion of a more benevolent 'social debt'. Here there is a little story told, a kind of myth. We are all born with an infinite debt to the society that raised, nurtured, fed and clothed us, to those long dead who invented our language and traditions, to all those who made it possible for us to exist. In ancient times we thought we owed this to the gods (it was repaid in sacrifice, or, sacrifice was really just the payment of interest – ultimately, it was repaid by death). Later the debt was adopted by the state, itself a divine institution, with taxes substituted for sacrifice, and military service for one's debt of life. Money is simply the concrete form of this social debt, the way that it is managed. Keynesians like this sort of logic. So do various strains of socialist, social democrats, even crypto-fascists like Auguste Comte (the first, as far as I am aware, to actually coin the phrase 'social debt'). But the logic also runs through much of our common sense: consider for instance, the phrase, 'to pay one's debt to society', or, 'I felt I owed something to my country', or, 'I wanted to give something back.' Always, in such cases, mutual rights and obligations, mutual commitments – the kind of relations that genuinely free people could make with one another – tend to be subsumed into a conception of 'society' where we are all equal only as absolute debtors before the (now invisible) figure of the king, who stands in for your mother, and by extension, humanity.

What I am suggesting, then, is that while the claims of the impersonal market and the claims of 'society' are often juxtaposed – and certainly have had a tendency to jockey back and forth in all sorts of practical ways – they are both ultimately founded on a very similar logic of violence. Neither is this a mere matter of historical origins that can be brushed away as inconsequential: neither states nor markets can exist without the constant threat of force.

One might ask, then, what is the alternative?

Towards a History of Virtual Money

Here I can return to my original point: that money did not originally appear in this cold, metal, impersonal form. It originally appears in the form of a measure, an abstraction, but also as a relation (of debt and obligation) between human beings. It is important to note that historically it is commodity money that has always been most directly linked to violence. As one historian put it, 'bullion is the accessory of war, and not of peaceful trade.'[1]

The reason is simple. Commodity money, particularly in

the form of gold and silver, is distinguished from credit money most of all by one spectacular feature: it can be stolen. Since an ingot of gold or silver is an object without a pedigree, throughout much of history bullion has served the same role as the contemporary drug dealer's suitcase full of dollar bills, as an object without a history that will be accepted in exchange for other valuables just about anywhere, with no questions asked. As a result, one can see the last 5,000 years of human history as the history of a kind of alternation. Credit systems seem to arise, and to become dominant, in periods of relative social peace, across networks of trust, whether created by states or, in most periods, transnational institutions, whilst precious metals replace them in periods characterised by widespread plunder. Predatory lending systems certainly exist at every period, but they seem to have had the most damaging effects in periods when money was most easily convertible into cash.

> **neither states nor markets can exist without the constant threat of force**

So as a starting point to any attempt to discern the great rhythms that define the current historical moment, let me propose the following breakdown of Eurasian history according to the alternation between periods of virtual and metal money:

I. Age of the First Agrarian Empires (3500-800 BCE)
Dominant money form: virtual credit money

Our best information on the origins of money goes back to ancient Mesopotamia, but there seems no particular reason to believe matters were radically different in Pharaonic Egypt, Bronze Age China, or the Indus Valley. The Mesopotamian economy was dominated by large public institutions (Temples and Palaces) whose bureaucratic administrators effectively created money of account by establishing a fixed equivalent between silver and the staple crop, barley. Debts were calculated in silver, but silver was rarely used in transactions. Instead, payments were made in barley or in anything else that happened to be handy and acceptable. Major debts were recorded on cuneiform tablets kept as sureties by both parties to the transaction.

Certainly, markets did exist. Prices of certain commodities that were not produced within Temple or Palace holdings, and thus not subject to administered price schedules, would tend to fluctuate according to the vagaries of supply and demand. But most actual acts of everyday buying and selling, particularly those that were not carried out between absolute strangers, appear to have been made on credit. 'Ale women', or local innkeepers, served beer, for example, and often rented rooms; cus-

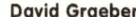
David Graeber

tomers ran up a tab; normally, the full sum was dispatched at harvest time. Market vendors presumably acted as they do in small scale markets in Africa, or Central Asia, today, building up lists of trustworthy clients to whom they could extend credit.

The habit of money at interest also originates in Sumer – it remained unknown, for example, in Egypt. Interest rates, fixed at 20 percent, remained stable for 2,000 years. (This was not a sign of government control of the market: at this stage, institutions like this were what made markets possible.) This, however, led to some serious social problems. In years with bad harvests especially, peasants would start becoming hopelessly indebted to the rich, and would have to surrender their farms and, ultimately, family members, in debt bondage. Gradually, this condition seems to have come to a social crisis – not so much leading to popular uprisings, but to common people abandoning the cities and settled territory entirely and becoming semi-nomadic 'bandits' and raiders. It soon became traditional for each new ruler to wipe the slate clean, cancel all debts, and declare a general amnesty or 'freedom', so that all bonded labourers could return to their families. (It is significant here that the first word for 'freedom' known in any human language, the Sumerian *amarga*, literally means 'return to mother'.) Biblical prophets instituted a similar custom, the Jubilee, whereby after seven years all debts were similarly cancelled. This is the direct ancestor of the New Testament notion of 'redemption'. As economist Michael Hudson has pointed out, it seems one of the misfortunes of world history that the institution of lending money at interest disseminated out of Mesopotamia without, for the most part, being accompanied by its original checks and balances.

II. Axial Age (800 BCE – 600 CE)
Dominant money form: coinage and metal bullion

This was the age that saw the emergence of coinage, as well as the birth, in China, India and the Middle East, of all major world religions.[2] From the Warring States period in China, to fragmentation in India, and to the carnage and mass enslavement that accompanied the expansion (and later, dissolution) of the Roman Empire, it was a period of spectacular creativity throughout most of the world, but of almost equally spectacular violence.

Coinage, which allowed for the actual use of gold and silver as a medium of exchange, also made possible the creation of markets in the now more familiar, impersonal sense of the term. Precious metals were also far more appropriate for an age of generalised warfare, for the obvious reason that they could be stolen. Coinage, certainly, was not invented to facilitate trade (the Phoenicians, consummate traders of the ancient world, were among the last to adopt it). It appears to have been first invented to pay soldiers, probably first of all by rulers of Lydia in Asia Minor to pay their Greek mercenaries. Carthage, another great trading nation, only started minting coins very late, and then explicitly to pay its foreign soldiers.

Throughout antiquity one can continue to speak of what Geoffrey Ingham has dubbed the 'military-coinage complex'. He may have been better to call it a 'military-coinage-slavery complex', since the diffusion of new military technologies (Greek hoplites, Roman legions) was always closely tied to the capture and marketing of slaves. The other major source of slaves was debt: now that states no longer periodically wiped the slates clean, those not lucky enough to be citizens of the major military city-states – who were generally protected from predatory lenders – were fair game. The credit systems of the Near East did not crumble under commercial competition; they were destroyed by Alexander's armies – armies that required half a ton of silver bullion per day in wages. The mines where the bullion was produced were generally worked by slaves. Military campaigns in turn ensured an endless flow of new slaves. Imperial tax systems, as noted, were largely designed to force their subjects to create markets, so that soldiers (and also, of course, government officials) would be able to use that bullion to buy anything they wanted. The kind of impersonal markets that once tended to spring up between societies, or at the fringes of military operations, now began to permeate society as a whole.

However tawdry their origins, the creation of new media of exchange – coinage appeared almost simultaneously in Greece, India, and China – appears to have had profound intellectual effects. Some have even gone so far as to argue that Greek philosophy was itself made possible by conceptual innovations introduced by coinage. The most remarkable pattern, though, is the emergence, in almost the exact times and places where one also sees the early spread of coinage, of what were to become modern world religions: prophetic Judaism, Christianity, Buddhism, Jainism, Confucianism, Taoism, and eventually, Islam. While the precise links are yet to be fully explored, in certain ways, these religions appear to have arisen in direct reaction to the logic of the market. To put the matter somewhat crudely: if one relegates a certain social space simply to the selfish acquisition of material things, it is almost inevitable that soon someone else will come to set aside another domain in which to preach that, from the perspective of ultimate values, material things are unimportant, and selfishness – or even the self – illusory.

III. The Middle Ages (600 CE – 1500 CE)[3]
The return to virtual credit money

If the Axial Age saw the emergence of complementary ideals of commodity markets and universal world religions, the Middle Ages were the period in which those two institutions began to merge. Religions began to take over the market systems. Everything from international trade to the organisation of local fairs increasingly came to be carried out through social networks defined and regulated by religious authorities. This enabled, in turn, the return throughout Eurasia of various forms of virtual credit money.

In Europe, where all this took place under the aegis of Christendom, coinage was only sporadically, and unevenly, available. Prices after 800 AD were calculated largely in terms of an old Carolingian currency that no longer existed (it was actually referred to at the time as 'imaginary money'), but ordinary day-to-day buying and selling was carried out mainly through other means. One common expedient, for example, was the use of tally-sticks, notched pieces of wood that were broken in two as records of debt, with half being kept by the creditor, half by the debtor. Such tally-sticks were still in common use in much of England well into the 16th century. Larger transactions were handled through bills of exchange, with the great commercial fairs serving as their clearing houses. The Church, meanwhile, provided a legal framework, enforcing strict controls on the lending of money at interest and prohibitions on debt bondage.

> **Most of the Medieval period saw money largely delinked from coercive institutions**

The real nerve centre of the Medieval world economy, though, was the Indian Ocean, which along with the Central Asia caravan routes connected the great civilisations of India, China, and the Middle East. Here, trade was conducted through the framework of Islam, which not only provided a legal structure highly conducive to mercantile activities (while absolutely forbidding the lending of money at interest), but allowed for peaceful relations between merchants over a remarkably large part of the globe, allowing the creation of a variety of sophisticated credit instruments. Actually, Western Europe was, as in so many things, a relative late-comer in this regard: most of the financial innovations that reached Italy and France in the 11th and 12th centuries had been in common use in Egypt or Iraq since the 8th or 9th centuries. The word 'cheque', for example, derives from the Arab sakk, and appeared in English only around 1220 AD.

> **virtual money is not necessarily an effect of capitalism, it might mean its direct opposite**

The case of China is even more complicated: the Middle Ages there began with the rapid spread of Buddhism, which, while it was in no position to enact laws or regulate commerce, did quickly move against local usurers by its invention of the pawn shop – the first pawn shops being based in Buddhist temples as a way of offering poor farmers an alternative to the local usurer. Before long, though, the state reasserted itself, as the state always tends to do in China. But as it did so, it not only regulated interest rates and attempted to abolish debt peonage, it moved away from bullion entirely by inventing paper money. All this was accompanied by the development, again, of a variety of complex financial instruments.

All this is not to say that this period did not see its share of carnage and plunder (particularly during the great nomadic invasions) or that coinage was not, in many times and places, an important medium of exchange. Still, what really characterises the period appears to be a movement in the other direction. Most of the Medieval period saw money largely delinked from coercive institutions. Money changers, one might say, were invited back into the temples, where they could be monitored. The result was a flowering of institutions premised on a much higher degree of social trust.

IV. Age of European Empires (1500-1971)
The return of precious metals

With the advent of the great European empires – Iberian, then North Atlantic – the world saw both a reversion to mass enslavement, plunder, and wars of destruction, and the consequent rapid return of gold and silver bullion as the main form of currency. Historical investigation will probably end up demonstrating that the origins of these transformations were more complicated than we ordinarily assume. Some of this was beginning to happen even before the conquest of the New World. One of the main

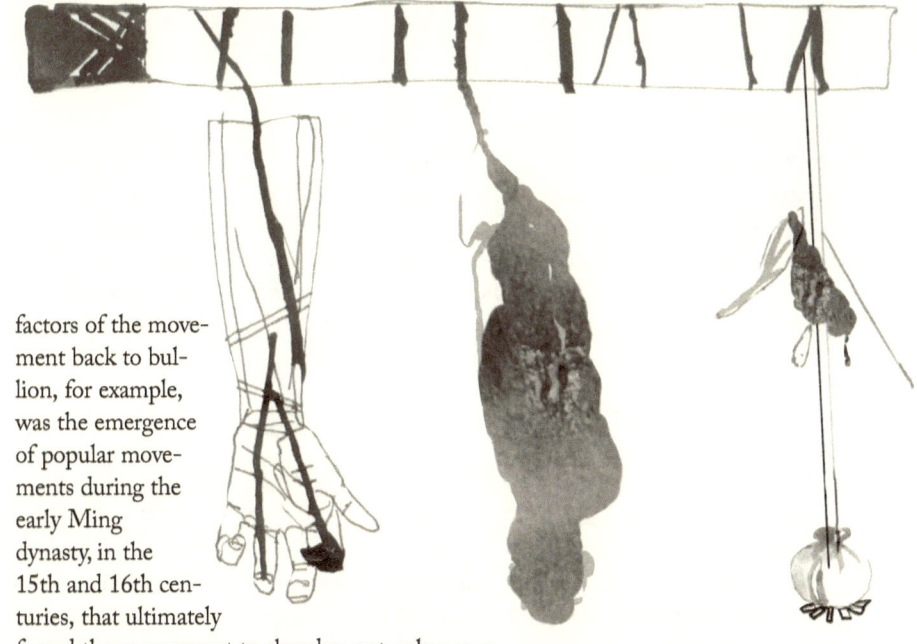

factors of the movement back to bullion, for example, was the emergence of popular movements during the early Ming dynasty, in the 15th and 16th centuries, that ultimately forced the government to abandon not only paper money but any attempt to impose its own currency. This led to the reversion of the vast Chinese market to an uncoined silver standard. Since taxes were also gradually commuted into silver, it soon became the more or less official Chinese policy to try to bring as much silver into the country as possible, so as to keep taxes low and prevent new outbreaks of social unrest. The sudden enormous demand for silver had effects across the globe. Most of the precious metals looted by the conquistadors and later extracted by the Spanish from the mines of Mexico and Potosi (at almost unimaginable cost in human lives) ended up in China. These global scale connections that eventually developed across the Atlantic, Pacific, and Indian Oceans have of course been documented in great detail. The crucial point is that the delinking of money from religious institutions, and its relinking with coercive ones (especially the state), was here accompanied by an ideological reversion to 'metallism'.[4]

Credit, in this context, was on the whole an affair of states that were themselves run largely by deficit financing, a form of credit which was, in turn, invented to finance increasingly expensive wars. Internationally the British Empire was steadfast in maintaining the gold standard through the 19th and early 20th centuries, and great political battles were fought in the United States over whether the gold or silver standard should prevail.

This was also, obviously, the period of the rise of capitalism, the industrial revolution, representative democracy, and so on. What I am trying to do here is not to deny their importance, but to provide a framework for seeing such familiar events in a less familiar context. It makes it easier, for instance, to detect the ties between war, capitalism, and slavery. The institution of wage labour, for instance, has historically emerged from within that of slavery (the earliest wage contracts we know of, from Greece to the Malay city states, were actually slave rentals), and it has also tended, historically, to be intimately tied to various forms of debt peonage – as indeed it remains today. The fact that we have cast such institutions in a language of freedom does not mean that what we now think of as economic freedom does not ultimately rest on a logic that has for most of human history been considered the very essence of slavery.

V. Current Era (1971 onwards)
The empire of debt

The current era might be said to have been initiated on 15 August 1971, when US President Richard Nixon officially suspended the convertibility of the dollar into gold and effectively created the current floating currency regimes. We have returned, at any rate, to an age of virtual money, in which consumer purchases in wealthy countries rarely involve even paper money, and national economies are driven largely by consumer debt. It's in this context that we can talk about the 'financialisation' of capital, whereby speculation in currencies and financial instruments becomes a domain unto itself, detached from any immediate relation with production or even commerce. This is of course the sector that has entered into crisis today.

What can we say for certain about this new era? So far, very, very little. Thirty or forty years is nothing in terms of the scale we have been dealing with. Clearly, this period has only just begun. Still, the foregoing analysis, however crude, does allow us to begin to make some informed suggestions.

Historically, as we have seen, ages of virtual, credit money have also involved creating some sort of overarching institutions – Mesopotamian sacred kingship, Mosaic jubilees, Sharia or Canon Law – that place some sort of controls on the potentially catastrophic social consequences of debt. Almost invariably, they involve institutions (usually not strictly coincident to the state, usually larger) to protect debtors. So far the movement this time has been the other way around: starting with the '80s we have begun to see the creation of the first effective planetary administrative system, operating through the IMF, World Bank, corporations and other financial institutions, largely in order to protect the interests of creditors. However, this apparatus was very quickly thrown into crisis, first by the very rapid development of global social movements (the alter-globalisation movement), which effectively destroyed the moral authority of institutions like the IMF and left many of them very close to bankrupt, and now by the current banking crisis and global economic collapse. While the new age of virtual money has only just begun and the long term consequences are as yet entirely unclear, we can already say one or two things. The first is that a movement towards virtual money is not in itself, necessarily, an insidious effect of capitalism. In fact, it might well mean exactly the opposite. For much of human history, systems of virtual money were designed and regulated to ensure that nothing like capitalism could ever emerge to begin with – at least not as it appears in its present form, with most of the world's population placed in a condition that would in many other periods of history be considered tantamount to slavery. The second point is to underline the absolutely crucial role of violence in defining the very terms by which we imagine both 'society' and 'markets' – in fact, many of our most elementary ideas of freedom. A world less entirely pervaded by violence would rapidly begin to develop other institutions. Finally, thinking about debt outside the twin intellectual straitjackets of state and market opens up exciting possibilities. For instance, we can ask: in a society in which that foundation of violence had finally been yanked away, what exactly would free men and women owe each other? What sort of promises and commitments should they make to each other?

Let us hope that everyone will someday be in a position to start asking such questions. At times like this, you never know.

Footnotes

1 Geoffrey W. Gardiner, 'The Primacy of Trade Debts in the Development of Money', in Randall Wray (ed.), *Credit and State Theories of Money: The Contributions of A. Mitchell Innes*, Cheltenham: Elgar, 2004, p.134.

2 The phrase the 'Axial Age' was originally coined by Karl Jaspers to describe the relatively brief period between 800 BCE – 200 BCE in which, he believed, just about all the main philosophical traditions we are familiar with today arose simultaneously in China, India, and the Eastern Mediterranean. Here, I am using it in Lewis Mumford's more expansive use of the term as the period that saw the birth of all existing world religions, stretching roughly from the time of Zoroaster to that of Mohammed.

3 I am here relegating most of what is generally referred to as the 'Dark Ages' in Europe into the earlier period, characterised by predatory militarism and the consequent importance of bullion: the Viking raids, and the famous extraction of *danegeld* from England in the 800s, might be seen as one the last manifestations of an age where predatory militarism went hand and hand with hoards of gold and silver bullion.

4 The myth of barter and commodity theories of money was of course developed in this period.

David Graeber <d.graeber@gold.ac.uk> undertook his original research in the relations between former nobles and former slaves in a rural community in Madagascar; it was about magic as a tool of politics, about the nature of power, character, and the meaning of history. He has recently completed a research project on social movements dedicated to principles of direct democracy and has written widely on the relation between anthropology and anarchism. He is currently working on a project about the history of debt

HUNGRY GHOST

Steve McQueen's controversial film about Irish republican hunger striker, Bobby Sands, eschews an articulation of politics, dramatising, instead, the body's capacity to 'speak'. But this microscopic focus on the flesh leaves us hungry for the political passions that animated it, writes Paul Helliwell

> *Words are shit, because they put you somewhere else. I'm trying to catch the things that are in between.*
>
> – Steve McQueen

A poll of critics in *Sight and Sound* (January 2009) has hailed *Hunger* as the best movie of 2008 and the arrival of Steve McQueen as a serious film-maker. It is an example of the film's surprisingly smooth reception in the UK and Ireland, given that it is based on the deaths of ten men on hunger strike in a British prison; men who had taken up arms against the British State. But it has not been uniformly smooth, with David Cox in his *Guardian* blog intent on reopening the wounds of the time: 'Far from being shocked at seeing the inmates roughed up a bit, I found myself wishing they'd been properly tortured.'[1] Who? The actors? There is a confusion of film and reality here. A desire to show the differences between now and then. Let us pursue them.

Re-enactment in politics often affirms the closure of questions, fixing meanings. In art or film it often serves to re-open questions, to bring something back. In both there is a faith that re-enactment is not numbing repetition but rehearsal, practice making perfect.[2] Think of the Northern Ireland marching season as an attempt to legitimise protestant ascendancy, to prevent the redistribution of roles within Ulster. But what are the aesthetics and the politics of McQueen's re-enactment, narrative and *mise-en-scène*? What is the redistribution being proposed here? What is McQueen trying to bring back?

'McQueen and his collaborators take us to a time and place that already seems unimaginable', says Ian Christie in *Sight and Sound*. To be honest, it is not just the worryingly accurate '70s and '80s clothes, breakfasts, gender roles, suburban repression, and concrete that seem unimaginable, but the degree of political conviction of those years. The end titles tell us the hunger strike ended with the recognition of every demand but the key one – that republican prisoners be recognised as political prisoners and treated as a special category. Just as the war in the six coun-

Paul Helliwell

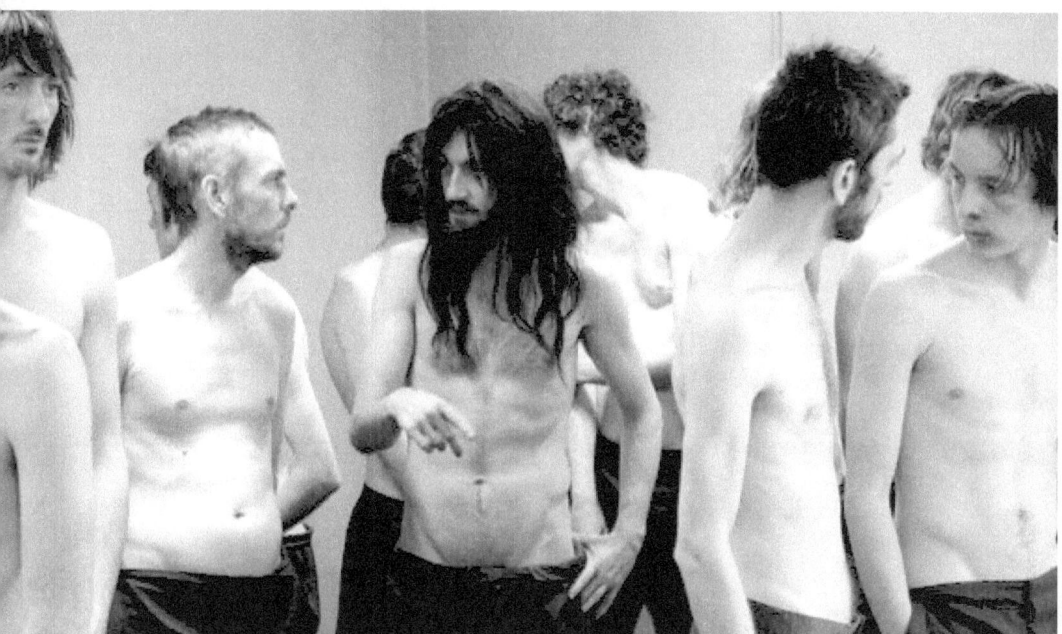

ties (or 'the troubles', if you will) ended with a recognition of almost every demand but the key one – that of a united Ireland. The 'no-state' solution is simply not recognisable as the object of the prisoners' struggle. As Bobby Sands' sister Bernadette Sands-McKevitt put it, 'Bobby did not die for cross-border bodies with executive powers'. Neither is it recognisable as the object of the struggle of the wider republican movement, nor indeed that of the loyalists. This is not the ending anybody imagined at the time. But does this really make that time unimaginable now? As Vikki Bell notes in her study of the Civic Forum, there the past and its politics must be dealt with carefully to prevent a collapse of civil society back into sectarianism. The peace remains haunted.[3]

There can be few who actively want to bring back 'the troubles' – so what is it McQueen wants to bring back? As he has repeatedly emphasised in interviews, what his 11-year-old self found inspiring when he first saw Bobby Sands on the news, and the artist of the present continues to find inspiring in Northern Ireland, is *the idealism* shown by all sides. In the film it is this abstract idealism that drives the violence, displacing politics from its rightful place.[4] This displacement is deliberate. For McQueen, the film is 'steeped in politics', however, he continues, 'politicians make a situation, but I'm interested in how people deal with that situation.' Robin Gutch, the producer of the film, added, 'I think what Steve and his team have done is restore the humanity to something that was clouded by a lot of ideology and rhetoric and posturing on all sides.'[5]

All images: stills from Steve McQueen's *Hunger*, 2008

Hungry Ghost

It is a brave decision to tackle this subject, but if there is one consistent criticism of films that have previously used 'the troubles' as material it is their repeated depoliticisation of peoples' motivations.[6] Foucault, commenting on French films, said that to make de Gaulle appear the saviour of the nation required hiding the extent of the resistance to the Germans: 'people are shown not what they were but what they must remember having been.'[7]

In *Hunger*, this de-politicisation is achieved by the additive nature of events on screen (and then..., not, and so...), mere practices previewed, echoed and rehearsed until they become static, embedded in epic rather than narrative time. Like the journey of Ulysses, we all know how this is going to end, but the events do not help us get there. Do the messages passed by the prisoners to their visitors cause the murder of the prison guard? We must decide, for we are neither shown nor told. It is a relief not to be hustled along by narrative, by plot mechanics, but while this collage of events (like the new Northern Ireland) has enough space for everybody, it has no room for their motivations. These we fill in as we choose. The width of McQueen's frame allows him to show the grain of concrete and of human flesh close up, but also to stage some egregious pieties: a riot policeman's cries (while his colleagues get on with the beatings), and a prison hospital orderly with 'UDA' on his knuckles carrying a dying Bobby Sands. Ideal reconciliations – we are all human, we all have ideals, and we all share the frame together.

In a film called *Hunger*, the gnawing of physical hunger is absent. In a filmed interview, McQueen explains that what attracted him to the story was the idea of not eating to make your voice louder, as he makes a gesture of gathering the food to his lips and then releasing it as speech. This is a reading similar to those of anorexia which understand it as an attempt to gain control of orality by refusing to eat. To his credit, McQueen uses it not as a means of *explaining away* the hunger strikers' actions but as a practice.[8] For McQueen is interested in repeatable practices, an archaeology of limit experiences. He wants us to see and experience 'something you cannot find in books and archives', 'the actual shit of history'.[9] This is McQueen's hunger. He is the hungry ghost, returning to eat, hungry for experience, and this is why the film, having brought us twice to the H-Blocks, keeps us there almost entirely until the end.

This is a movie of bloody knuckles and suppurating bed sores, but also of shit smeared on the walls and mounds of rotting food refuse in the cells. For the practice McQueen is really interested in is not the hunger strike, but the blanket and dirty protests – the refusal to wear prison uniform and slop out the excrement from the cell – shown in the first third of the movie. Here the shots inside the prison that do not feature the suffering human body show the dirt and abjection in which the prisoners lived for four and a half years – in one long, fixed-position shot, piss flows from

Like the journey of Ulysses, we know how this will end, but events don't help us get there

under closed cell doors before it is disinfected and swept back in by a masked and rubber suited prison officer. The piss is caused to flow out under the doors by means of a dam of festering food remains (we know this because we have been shown it earlier but now the men doing it remain hidden). In the unlikely event that this scene could have been filmed in the '70s, this would probably have been portrayed as resistance and prank. Now we see these as if they were aesthetic practices documented on video and shown in an art gallery.[10] However visceral the violence or vile the filth, it is all form because we are at an aesthetic distance from its motivations. As Fintan O'Toole notes, the blanket men have become a kind of wish fulfilment of the avant-garde artist.[11]

For McQueen, the filth is fertile. It is the revolutionary city (of Bobby Sands), to be counterposed with the cleanliness and psychoanalytic repression of the suburbs that motivates the prison guards to clean and clean again, and the untrustworthy, false paradise of the countryside (where the priest comes from). It is an alchemical purification towards an ideal through corruption. When McQueen says it's a surreal situation, he means it not in the sense of being improbable but that the Surrealists would recognise it: 'In history as in nature, the rotten is the laboratory of life.' Then as now the question, disputed between Bataille and Breton, is: can the rotten suffice instead of politics?[12]

What were the politics of these practices that have been displaced by both history and McQueen? First we need a little context, but the beginnings can be made using Foucault's work on punishment and prisons in *Discipline and Punish*.[13]

For the British State, in 1975, it was necessary that the six county 'statelet' of Northern Ireland be normalised as part of the United Kingdom with (to all appearances) the same laws and penal policies as the 'mainland'. Moving it from a colonial war to a policy of 'Ulsterisation and Criminalisation', new IRA prisoners had to be jailed as criminals. This replaced the colonial policy of internment without trial, and hid the Emergency Powers Act and no-trial-by-jury Diplock courts. From the prisoners' side, their main aim was to challenge the right of the State of Northern Ireland to jail them by refusing to be produced as criminals by their imprisonment and to assert military discipline. This rejection of the prison is an attack on the legitimacy of the state as deriving not from an abstract 'above', but from its institutions, an attack on its sufficiency.

> If the crowd gathered round the scaffold, it was not simply to witness the sufferings of the condemned man [...], it was to hear an individual who had nothing more to lose curse the judges, the laws, the governments and the religion [...]. Under the protection of imminent death.[14]

Foucault argued that with the prison, punishment dare not speak its own name and instead becomes correction or cure. Instead of the condemned being tortured to death in public, they are instantaneously dispatched by lethal injection behind high walls, as if the judgement at public trial were sufficient punishment and the judicial murder a necessary but troubling detail. Prisons are run to timetables with adequate food, sanitation, and clothing – beatings and physical pain their dirty secret. These refusals by the prisoners to wear prison uniform, to slop out their excrement, and finally to eat were a contestation of the prison's logic of hiding and eliminating society's waste people. An elimination blocked, in microcosm, within each cell.

However visceral the violence or vile the filth, we are at an aesthetic distance from its motivations

Prisoners and their supporters smuggled out their accounts ('comms' – another practice shown at length in the film) and ensured their publication.[15] What this brought back, however, was a public punishment of the body of the condemned that had been replaced all over Europe in the early 19th century. As Foucault notes, the problem with the spectacular punishment of the body was that 'it was always ready to invert the shame visited on the victim into pity or glory' – from Jesus to Jesse James, these 'ambiguous rituals' invert into martyrology. Prison was to stifle words spoken 'under the protection of imminent death' by secluding prisoners and by hiding the 'excesses' and survivals of the earlier regime. Yet the earlier regime remains something that can be re-invoked by the state and its citizens equally; a demand that 'the punishment should fit the crime' exactly, so as to equal and erase it – an eye for an eye.[16] In *Hunger*, the beatings are preceded by the removal of the prisoners' record cards from the slot outside their cells, but this merely reveals the limits of the protection

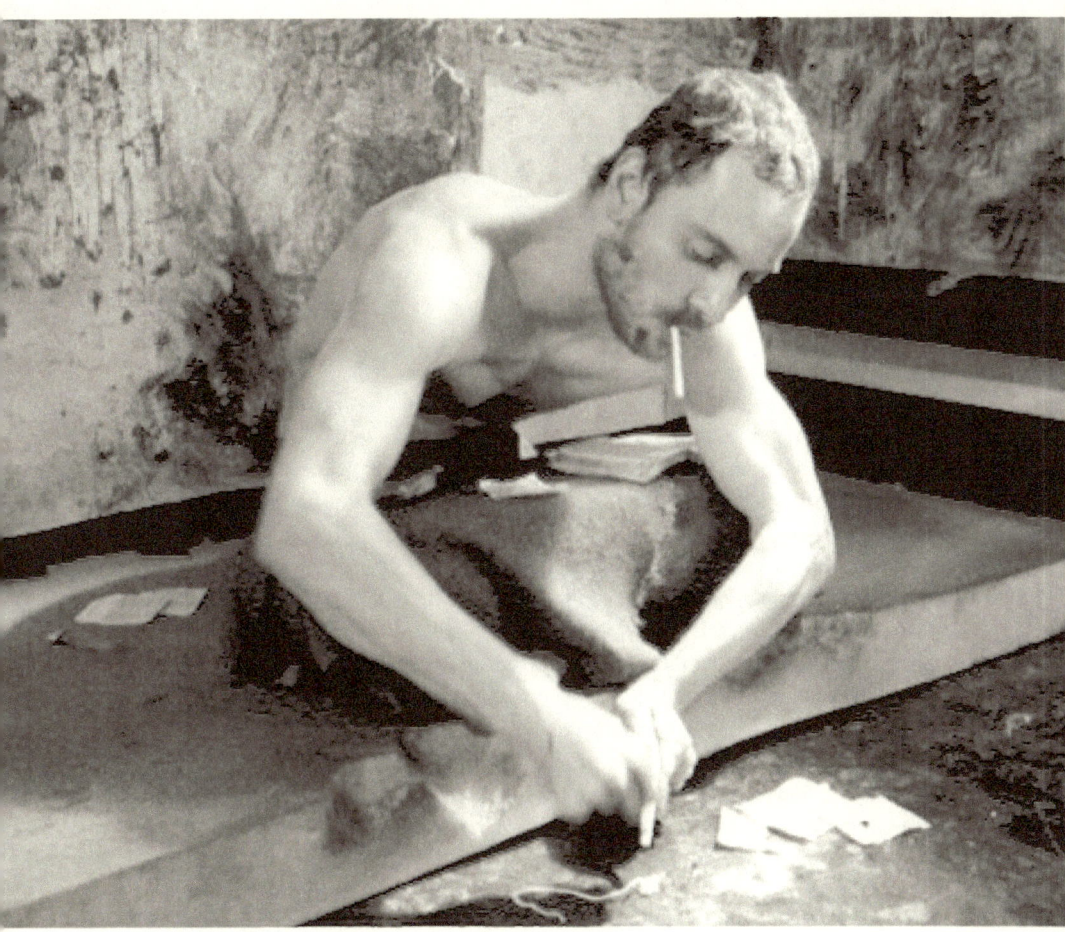

offered by the prison. Now we are often reduced to calling for the reinstatement of that protection as torture and ghost prisons have returned, as the footage of the hanging of a tyrant 'leaks'.

It is important to distinguish the political logic that led the Provisional IRA to allow the hunger strike to go ahead – the messy, insufficient logic of events – from the necessary logic of the film's story. There is no conveniently cast Judas or Coward Robert Ford, so the film must provide a reason as to why Bobby Sands and the others chose a road that we know will lead to their deaths – it is the central problem, the one thing in the movie that requires a cause. McQueen resolves it through a long, exhaustively rehearsed, nearly single-shot take between Sands and a Priest. It is difficult to tell what the film would have been if he had followed his original intention to have it without dialogue throughout; then again, he has also been reported as saying this scene was 'always there'. (On DVD we can, perhaps, try the film without it.) Film critic Demetrios Matheou says it's a scene 'to rival that of Pacino and De Niro in *Heat*'.[17] But that is the least of its problems.

Structurally, the film must provide Bobby Sands with a story that makes his death necessary. Here's the spoiler. A foal with a broken back leg is floundering in a stream in a forest, suffering, getting cut up by the rocks (this is when Sands is 11 or so, he's there as part of a cross country running team). He shoulders the responsibility, drowns the foal and takes the blame because he knows he's strong enough to bear it. I can see this scene clearly in my mind, but in the film it's only talked about. From then on we are committed to humanely killing the foal, to political struggle; we are prepared to accept the necessity of Sands' death. In the film's nearly wordless last third it all goes a bit *2001: A Space Odyssey*. As he sickens, and again, as he dies, Sands has a vision of himself as the 11-year-old runner. If indeed it hadn't earlier – the spiral of shit a prisoner daubs on the wall is accorded the same cinematic reverence as the black monolith. Transcendence is hard to show.[18]

And as I read it back I find this shaggy pony story unsatisfying. The script has incorporated the story of the killing of the foal as a moment of realisation, one that is much more powerful than the attempted elucidation of the prisoners' political situation in respect of the republican movement and the British government immediately prior to it in the scene. To use the stock phrase 'it was then that I realised', I had been moved by the kind of journey storytelling politicians think more effective than 'hard' facts. I'm not thinking Pacino and De Niro here, I'm thinking of Tony Blair telling us of meeting 'Sierra Man' while canvassing and realising at that moment that Labour needed to change to become electable. Everyone likes a story and we are not averse to accepting something as a fact because it seems it should be true.[19]

In his *Frieze* magazine review 'Flesh Becomes Words', Caoimhín Mac Giolla Léith maintains that McQueen distrusts language and instead favours the visceral qualities of sound.[20] Mac Giolla Léith recognises the role of a performative storytelling in *Hunger* and two of McQueen's earlier artworks (*7th November* and *Girl's Tricky*), but he holds that these stories are undermined, made insufficient, and that the failure of the dialogue between the priest and Sands to prevent the hunger strike is an example of this distrust. (This is a good story but we already know the hunger strike went ahead – this only works if McQueen is dramatising his own distrust of language, but he needs this scene to work.) He quotes McQueen as saying, 'Words are shit, because they put you somewhere else. I'm trying to catch the things that are in between.'

It is a philosophical commonplace that behind the fog of words is real direct experience. McQueen wishes to take us there and to experience it himself, just think of the conditions under which this film was made (in a prison, with starved actors and mounds of festering food refuse). If 'words (are shit)' is the bad in the opposition McQueen has constructed, his good is probably the body as a site of unmediated experience. But the true lesson of the folk stories in which Trickster, Hermes, Anansi or Raven find something valuable in a pile of shit (after searching for years and getting very dirty) is that another distribution is possible; a collapse of false opposites, such as sought by the Surrealists. What McQueen is searching for is to be found in his shit, in the 'ideology and rhetoric and posturing' he has discarded, in the politics he has divorced from people dealing with situations, but finally, in words.

> **men appear only as bodies, their words as silence. Everyone has been spoken for**

McQueen has confined the words to one scene, words necessary to prevent a 'misreading' of Sands' motivation. But Mac Giolla Léith takes this distrust further. The flesh the hunger strikers lost was transmuted into words, words laboriously smuggled out, but so many words that they are not worth reading. And these are what have really been lost from this film, these words and the politics and the movements that these words called into being, the flesh of these words. It is not that 'men and events appear [...] as shadows that have lost their bodies', but that men appear only as bodies, events as shadows, and their words as silence.[21] Everyone has been spoken for.

McQueen has spoken of the process of making the film itself not just in a manner reminiscent of relational art, but as if it was a therapeutic working through, as if art had better process than politics:

> When a person whose father or uncle or sister was involved in 'the troubles', and they can do something about it physically, involve themselves in the making of the film, then they feel empowered by the physicality of doing something within that.[22]

But there is a fear of re-enactment here also. Permission to shoot inside the H-Blocks themselves was refused so *Hunger* was shot elsewhere, ironically on sets based on the Baader-Meinhof prison (whether sets from the film *Baader Meinhof Complex*, or based on Stammheim prison or both is not made clear). McQueen views the denial of permission to film within the H-Blocks as a blessing because he was unsure of what 'atmosphere' it would have cast over the crew.[23]

McQueen has a topical hope that the film will help in the understanding of how 'idealism' makes martyrs. He could not be more wrong. It is not idealism that makes martyrs, but politics. He is afraid that the past will steal its scene, for the present that has reconfigured the past remains haunted by what is eliminated.

McQueen has made a film on 'the troubles' that manages to eliminate the presence of the British Army, leaving only Margaret Thatcher's disembodied voice. What would disappear from a film about the Gulf War?

The pointy hood, the electrodes, the prisoner who stands on a box. It is not a Goya – it is a human being tortured. Here art effects no redistribution but of itself.

Footnotes

1 David Cox, '*Hunger* strikes a very sour note', http://www.guardian.co.uk/film/filmblog/2008/nov/03/hunger-bobby-sands. Comments such as these generated over 700 postings within ten days before the tail was shut down. The adjudication by readers' editor Siobhan Butterworth itself generated a further 400 comments.

2 See Hillel Schwartz, 'Once More With Feeling', in *The Culture of the Copy*, New York, Zone Books, 1996, chapter 7. Artists have shown great interest in re-enactment; Melanie Gilligan has been the only one, to my knowledge, bucking the trend by being interested in pre-enactment, the ability of role play to tell the future, taking her cue from books such as Michael D Schrage's *Serious Play: How the World's Best Companies Simulate to Innovate*, Harvard Business School Press, 1999.

3 Foucault's concept of 'Governmentality' might be a useful guide here as to how a state can be reformed as just the sum of its administrative arrangements rather than as that from which sovereignty derives. See Vikki Bell, 'Spectres of Peace: Civic Participation in Northern Ireland', *Social and Legal Studies*, Vol.13, No.3, pp.403-428. The Civic Forum, composed of representatives from the business, trade union and voluntary sectors, was established as part of the Northern Ireland Act of 1998 to act as a consultative mechanism.

4 In *Sight and Sound*, Colin McCabe notes with approval that McQueen has 'navigated the politics of Northern Ireland by studiously ignoring them'; Ronan Bennett, an earlier prisoner at Long Kesh/the Maze has applauded his 'eschewing explication'; Maria Fusco, in *Art Monthly*, noted that politics was the 'one thing missing'. See 'Films of 2008', *Sight and Sound*, January 2009; Ronan Bennett in, http://www.guardian.co.uk/politics/2008/oct/22/maze-prison-film-northernireland-hunger

5 See, http://www.eyeforfilm.co.uk/feature.php?id=563

6 John Hill, 'Images of Violence', in Kevin Rockett, Luke Gibbons, John Hill, Croom Helm (eds.), *Cinema and Ireland*, Syracuse University Press, 1987, pp.147-193; and Ruth Barton, *Irish National Cinema*, Routledge, 2004, pp.157-78. Not to mention an extremely effective campaign of censorship of the news aimed at the same thing. See Liz Curtis, *Ireland and the Propaganda War*, Pluto Press, 1983.

7 Michel Foucault, 'Film and Popular Memory', in *Foucault Live*, Sylvère Lotringer (ed.), Semiotext(e), 1989, p.123.

8 See, http://www.filmcatcher.com/interview_detail/152/. Caoimhín Mac Giolla Léith draws a comparison with Maud Ellman's *The Hunger Artists: Starving, Writing and Imprisonment*, Virago Press, London 1993. The psychoanalytic reading is seen in an inverted Oedipal scene told as a joke by the prison officer, one spread over an additional generation. The punch line, 'See it's not so funny when it's your mammy', leaves the child Oedipus having the last laugh. If Oedipus is interpreted as challenging the father for language then this can be interpreted as keeping the right to speak open. This is arguably McQueen's aim.

9 Tony Rayns, 'Hunger', p.63; and Kieron Corless, 'BFI 52nd London Film Festival and Interviews', pp.24-27, both in *Sight and Sound*, November 2008.

10 This scene is something of an exception. Something of the prank still remains in that, with no human actors, what it most closely resembles is Fischli and Weiss, *The Way of Things*.

11 Fintan O'Toole, 'Hunger Fails to Wrest the Narrative From the Hunger Strikers', *Irish Times*, 22 November, 2008, http://www.irishtimes.com/newspaper/weekend/2008/1122/1227288132671.html

12 Alan S. Weiss, 'Between the Sign of the Scorpion and the Sign of the Cross: L'Age D'Or', in, R. Kuenzli (ed.), *Dada and Surrealist Film*, MIT, 1996, p.169. Most of this chapter can be read on Google Books. See also Paul Hammond (ed.), *The Shadow and its Shadow*, BFI, 1978.

13 The struggles of republican prisoners should be contrasted with the post-'68 struggle in France, where jailed Maoists abandoned their hunger strikes and demands for a separate political category and it became over prison itself. The IRA protests should be understood not as against prison itself, but functioning by means of prisons. See, http://www.espacestemps.net/document1164.html See also the second half of 'On Attica' in *Foucault Live*, Sylvère Lotringer (ed.), Semiotext(e), 1989.

14 Michel Foucault, *Discipline and Punish*, trans. Alan Sheridan, Vintage, 1979, pp.60-1.

15 These would perhaps provide a better basis for a Foucauldian study rather than the ad hoc use of Foucault's themes and examples that has been attempted before and that I repeat here. See Martin Wainright and Ben Quinn, 'The Past is So Last Year: New Archaeologists Dig the Present', *The Guardian*, 22 December, 2008.

16 The state abrogates the people's right to punish. The death threats which defence lawyers receive in high profile cases and David Cox's demands to see IRA members properly tortured arise because of this. Also, *Death Wish, Batman, Dirty Harry* etc. See 'Anxiety of Judging' in Lotringer op. cit.

17 Interestingly, *Heat* was a remake by Michael Mann of his own earlier TV movie *LA Takedown*. While it doesn't feature great actors, at least it doesn't have the scenes designed to produce great acting.

18 It also has a psychic twin in Carol Read's *Odd Man Out*, where Lukey the artist wants to paint Johnny, a wounded IRA man on the run, to catch him as he passes from one world into the next.

19 For a good discussion of this instrumentalisation of structuralism with examples see, http://news.bbc.co.uk/nol/shared/spl/hi/programmes/analysis/transcripts/21_02_08.txt

20 Caoimhín Mac Giolla Léith, 'Flesh Becomes Words', *Frieze*, September 2008, pp.125-131, http://www.frieze.com/issue/article/flesh_becomes_words

21 'Men and events appear as inverted schlemihls, as shadows that have lost their bodies', Karl Marx, '18th Brumaire of Louis Bonaparte', in *Marx and Engels: Selected Works*, Lawrence and Wishart, London, 1968, p.115.

22 Such claims are often made for re-enactments, e.g., for Paul Greengrass' *Bloody Sunday* (2002), Barton, op.cit., p.174.

23 See, http://www.channel4.com/film/reviews/feature.jsp?id=166630&page=2; http://www.iht.com/articles/2008/05/22/arts/fmmcqueen.php; http://www.filmcatcher.com/interview_detail/152/

Paul Helliwell <p.helliwell2000@yahoo.co.uk> would like to direct people to his blog on the myspace page of his 'brother ass' horsemouth: http://www.myspace.com/horsemouthfolk

A CLIMATIC DISORDER?

> At last November's NUM convened conference, trade unionists and Climate Campers were invited to debate the explosive cocktail of (clean) coal, class and climate change. **John Cunningham** reports on the frustrating attempts to find a middle ground

Last year's camp at Kingsnorth in Kent against the opening of a new coal fired power station produced a range of predictable responses, from the inanity of the *Guardian*'s suggestion that it was yet another alternative lifestyle festival, to the over zealous attentions of the security state. One of the more interesting responses was from long standing anarchist activist and ex-National Union of Mineworkers (NUM) official Dave Douglass. In a polemic against the camp, he addressed the anti-coal bias of the Climate Campers alongside a perceived lack of class analysis within the camp and the wider green movement. There is undoubtedly a feel good anti-capitalism implicit in much of the discourse around Climate Camp that can exclude any consideration of class in favour of blandly utopian sentiment. For instance, in the *Climate Camp Newspaper*, statements such as 'Sometimes it feels as though our world is coloured in sadness, and you just want to be somewhere else', read less like a *détournement* of advertising copy than a self help approach to political activism, the 'middle class voice' that Douglass characteristically claimed Climate Camp spoke in.[1]

At the time I was relieved that, against the fluffy anti-capitalism of much of the camp's official discourse, Douglass introduced the perspective of those who may not have 'somewhere else' to go, locked into jobs and communities that a politics of exodus cannot easily address. The yearly anti-climate change roadshow attempts to offer a response to climate change that would destabilise business as usual, suggesting at least nominally anti-capitalist alternatives. However, its model of protest camp and sustainable community gleaned from the post-Seattle summit protests can seem too abstracted from everyday life to break the general perception that climate change exists 'out there', to be dealt with by superheroes such as Al Gore. Its model of sustainability can also appear as a holiday in scarcity to the casual observer. The intervention by Douglass was a dose of messy actuality. The camp's response was to invite Douglass to address it, and he turned up with ex-NUM president, 'Old King Coal' himself, Arthur Scargill in tow. The Newcastle based conference, Class, Climate Change and Clean Coal – the Climate Campers and the Unions, sponsored by the NUM, the RMT and the Industrial Workers of the World (IWW), arose out of this dialogue.

> *This is a sad and confusing conjuncture of forces.*
> – Dave Douglass[2]

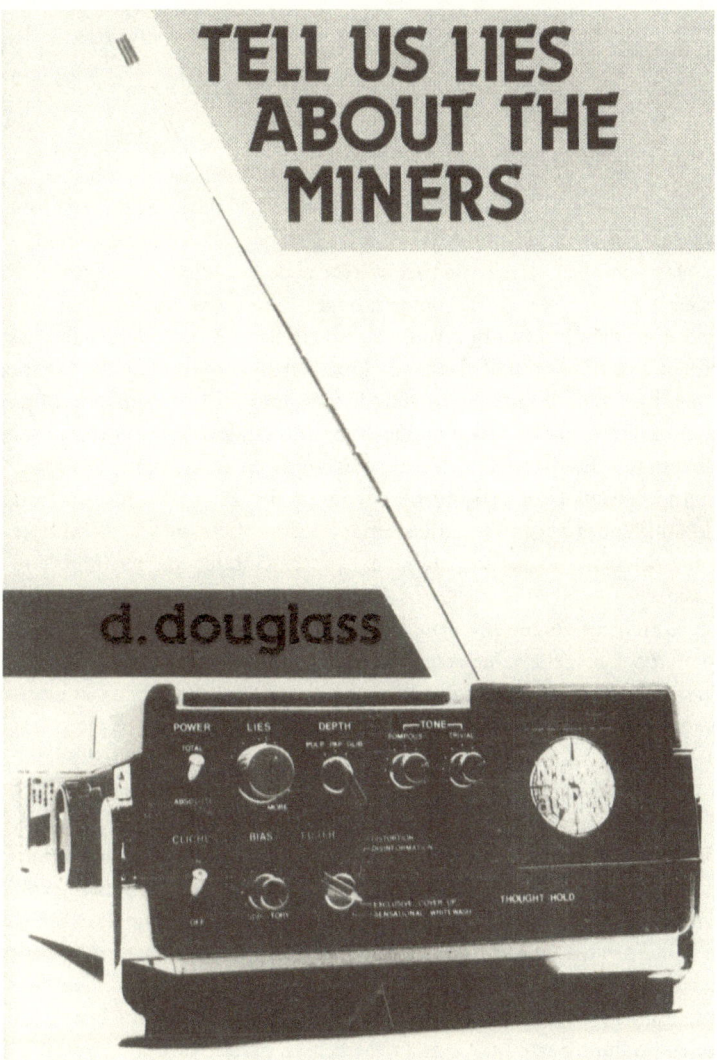

At least the intersection of labour and environmentalism in the Douglass/ NUM/ Climate Camp exchange punctured a certain spectacle of climate change: the accumulation of catastrophic images, millennial eco-fear and eco-friendly consumerism that can induce occasional dread and the desire to assume the crash position. The class tensions around Climate Camp seemed like a clash of cultures between a traditional 'mass worker' form of trade unionism and a diffuse network of activists whose politics ranged from pale green reformism to red and black anti-capitalism. In fairness to Climate Camp, they ran workshops on class and emphasised a 'just transition' in the official booklet's dialogue with workers in carbon-based industries. This is the notion that a transition can be made to a non-carbon based economy that does not penalise the poor or workers in carbon based industries such as coal miners. It is an

Image: Cover of Dave Douglass' 1987 pamphlet, *Tell Us Lies About the Miners: The Role of the Media in the Great Coal Strike of 1984/1985* argument for responses to climate change that place social justice at the forefront of any structural shift in the economy. While it is often posited as a decentralised, autonomous response, it can also be part of a social democratic state-led one, as Paul Chatterton's argument later in the conference for a 'green new deal' was to make clear.

While Douglass' intervention was welcome it also threatened to reduce a complex series of questions around class and climate change to the singular question of 'more prole than thou'. There's an admirable tenacity to the perpetuation of proletarian culture in the face of defeat and expropriation: many of the current pit banners at the miners' Durham Big Meeting gala are from defunct pits, less postmodern nostalgia than an assertion of community. However, this can harden into a closed identity as can the activist milieu around Climate Camp, with its own cultural and discursive forms and marginal counter-cultures that mean little to those outside.

I hoped that the conference would reveal something in common beyond a shared preference for renewable energy and being anti-nuclear, points all the delegates made. The main point of contention in terms of energy policy was clean coal, based on carbon dioxide capture and storage, but the conjunction of more anti-capitalist elements from Climate Camp with the representatives of old school trade unionism also suggested other fault lines, both cultural and political.[3]

Following the great tradition of working class radicals, the meeting was held in the Bridge Hotel pub in Newcastle, close to the industrial grandeur of the Tyne bridge and adjacent to the contemporary bubble of regeneration, cultural capital and service driven consumerism that is Newcastle today. A combination of NUM veterans, miners, trade unionists, eco-punks, socialists and maybe even one or two members of the general public attended the meeting. The conference delegates – NUM spokesmen, Climate Campers and a lone RMT spokesman – occupied a low wooden podium. Seated three at a time behind a desk, they sometimes gave the impression of a quarrelsome Stalinist tribunal. After a brief introduction by Keith Whittaker of the NUM, Dave Douglass presented the keynote speech.

> *The Earth disnae give a bugger.*
> – Dave Douglass

Douglass came across as humorous, thoughtful and angry, equally at home in both NUM and Climate Camp circles. In comparison to the other miners delegates his presentation was discursive and wide ranging. He admitted the human impact on climate change, but emphasised its natural movement, believing that natural factors are more likely to wipe out humanity, or even asteroids from space. In this he was ducking the issue since the issue is not so much whether climate change is 'natural' or 'man made' as the effect of a particular social relation, capitalism, on the way humanity interacts with the environment.[4]

He was on stronger ground when he argued that there was a reification of climate change that came out of a green view of nature: 'As a painted landscape pantomime at the Theatre Royal'. Nature, rather than being an abstraction, was always linked to the productive activity of humanity, the Earth as a 'dead crust' rather than some Gaia like entity. For myself, viewing the earth as a 'dead crust' ends up as much of a reification as environmentalist visions of Gaia. The productive activity that links humanity with nature is never simply exerted upon something inert – a 'dead crust' is as much of an abstraction as Gaia, both remaining within the logic of capital. The necessity was to address climate change from the perspective of the poor and exploited. Capitalism was incapable of doing so, but some trade unions like the NUM and IWW had this perspective. Renewables were fine but not wind farms, partly because they desecrated open spaces. He called for fair trade coal as a way of both invigorating the UK coal industry and addressing exploitation abroad. Douglass also emphasised the running down of the UK mining industry for political reasons, accusing the green movement of complicity with this process.

> The delegates sometimes gave the impression of a quarrelsome Stalinist tribunal

Environmentalism and Climate Camp undoubtedly attract a lot of green mystical cretinism, but Douglass' ire was misdirected in terms of the environmentalists present, all being scrupulously materialist in their thinking on climate change. Stressing the natural cycle of climate change meant that Douglass actually made more of Nature than they did.

Douglass also introduced a topic that other NUM speakers emphasised: the industrial working class, in this case miners, have an inherent political perspective and class consciousness lacking in the service industries. It's certainly impossible to imagine mass pickets of Pret A Manger workers confronting cops, although that would be quite an event. But it's equally impossible to (re)posit the mass worker as the vanguard of class consciousness, since the conditions for that kind of organisation no longer exist in the UK. Context changes all, and it is sobering to realise that whereas 10,000 miners confronted cops at Orgreave in 1984, there are now around 5,000 miners in total. A workerist ideology introduces nostalgia rather than facing the reality of disseminated struggles against capitalism, out of which elements of Climate Camp arise.

Other NUM delegates were more narrowly focused upon clean coal as a technological solution to climate change, the continued viability of carbon-based industry

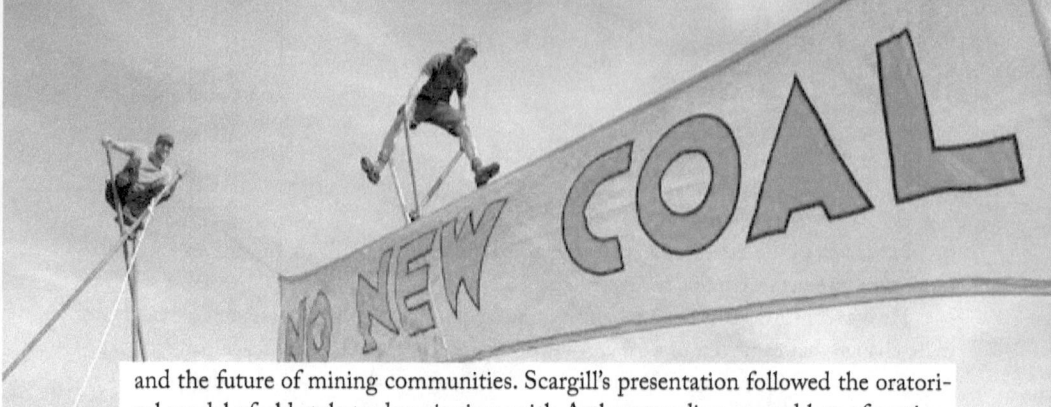

and the future of mining communities. Scargill's presentation followed the oratorical model of old style trade unionism, with Arthur standing up and lots of gesticulation and pointing. Speaking loudly he could have been addressing the Durham Big Meeting. He made a well researched, belligerent case for the continued relevance of coal to energy policy, packing his speech with statistics and questioning the emphasis upon coal as a cause of global warming, a point echoed by other NUM speakers. Scargill went on for twice the 20 minutes allotted, in true leftist style, with everyone from the chairman onwards probably too overawed/exhausted to intervene. He left soon after.

All of the Climate Campers were at pains to emphasise that they were not hostile to mining communities and were aware of the intrinsic relationship between climate change, class exploitation and capitalism. They also all underlined that they were not 'official' representatives of Climate Camp. This was undoubtedly one of the lines that separated the trade unionists from the Climate Campers, the union officials having a much more unproblematic relation to being a representative of the working class. The Climate Campers were definitely from the more anti-capitalist wing and it might have been interesting had someone from a more single-issue perspective been present. Paul Chatterton, activist and Leeds University academic, gave a well reasoned presentation about the need for a 'just transition'. After underlining the importance of avoiding a climate change 'tipping point' of a four degrees rise, he emphasised that environmentally based politics were ultimately against 'mindless, ceaseless growth' in the form of neoliberal capitalism. 'Just transition' would share out the costs of climate change equally, through a 'green new deal', ecological Keynesianism creating a 'green collar economy'. This would amount to the re-nationalisation of energy production and a rejection of the market.

I must admit that the concept of a 'green new deal' makes me want to strangle the planet with a couple of spare plastic bags. It's the realist corollary to the utopian elements of Climate Camp, but such an uncritical acceptance of a social democratic solution ignores the problem that capitalist social relations would still remain in place. It would be compatible with the development of an authoritarian, biopolitical state, obsessed with the administration of life. It is quite easy to imagine a dystopian 'green new deal' that continued the valorisation of capital alongside a work-ethic based morality all too conducive to the more sanctimonious elements of environmentalism. Chatterton did mention that a 'green new deal' might lead to less work and more holidays, a rare acknowledgement that climate change might not necessitate new regimes

Image: Climate Camp, 2008

John Cunningham

of scarcity. There is in this a trace of what was missing in the conference, a sense of possibility not embedded in soft focus 'somewhere else' utopianism but in an immanent engagement with capital's apparatus of capture. However, a 'green new deal' is unlikely to deliver the kind of simultaneous refusal of scarcity and production that might begin to construct a genuine anti-capitalist response to the exigencies of climate change. It hardly amounts to a critique of wage labour.

Ian Lavery, President of the NUM, underlined the gulf between the NUM and Climate Campers through his refusal to engage with Paul Chatterton's case for 'just transition'. Remarking dismissively that he was in the bar during Chatterton's talk, apparently what was needed was a 'just transition' to clean coal. Throughout the conference the NUM's concentration upon clean coal raised questions about the contradiction of trade unions being not only a bureaucratic appendage to the marketing of labour but also a possible focal point for resistance and the reproduction of communities tied to a particular industry. Lavery's work ethic was committed to coal rather than a green collar economy. He left shortly afterwards in his big car to go to another meeting. Oh, the life of the full time official.

> **the NUM official was resplendent in red shirt and tie, in case we didn't get the message**

The focus on new technologies as a general fix for climate change always threatens to introduce a Hollywood blockbuster narrative: 'And then there was clean coal...' While the viability of clean coal is in doubt, any present development of it is reliant upon capital being able to extract value from it.[5] The same would go for the development of renewables. It is unlikely that an exclusive focus on technology can really challenge the relation between climate change and the reproduction of capitalism.

David Guy, President of the North East area of the NUM, made a restrained and dignified argument for the viability of the North East Coalfield and spoke with some melancholy of the effects of the post-1984 strike on what used to be the 'left wing juggernaut' of the trade union movement. He pointed out that six months after the strike in 1985, and again in 1987, miners were back out on strike. A good reminder of how the memory of past struggle informs the present dilemma of the NUM, mining and ex-mining communities.

The next Climate Camper was Paul Morrozo, who made the connection between climate change and hurricane Katrina, pointing out that it was the precarious poor who suffered while the rich escaped, as demonstrated by the wholesale privatisation and gentrification of New Orleans. He also pointed out that climate change would put a squeeze on capitalism's attempts to avoid financial crisis and, I would add, its attempts to address climate change. More workers' control was essential to combat climate change. He was willing to countenance mining if carbon capture was developed as a transitional measure, but argued that it was still in an experimental phase; the state and E.on were both lying about Kingsnorth's potential for carbon capture. Ultimately, if resources continue to be extracted, 'we're toast'.

Morrozo's contribution was constantly interrupted by the chairman, NUM official Dave Hopper, resplendent in both red shirt and red tie in case we didn't get the message. Hopper seemed to take great offence at any anti-coal argument, cutting in at one point with 'Put one of those windmills on your head and walk around with it'. This might have been funny to him but most objected strongly, especially when he seemed to have a fit after being heckled by a woman. At this point he resigned to be replaced by Dave Douglass. This was one of the few points at which a residual animosity surfaced, the atmosphere generally being more constructive.

The RMT regional secretary Stan Herschel spoke on the influence of the road lobby and the confluence of interests in business that work against environmentally sustainable energy resources and the trade union movement. At this point, I must admit, my mind was drifting towards a pint and my own most sustainable way home.

The last speaker was Kevin Bland of Green Anarchism, who talked about the class nature of climate change as the poor carry the cost, a point NUM delegates had also made. He was against the continued mining of coal and questioned the environmental credentials of carbon capture. Open-cast mining was unacceptable and a form of revenge on mining communities. His description of the work done by local environmental groups against open-cast suggested to me a much more investigative and open process than much of the Climate Camp activists' grandstanding, since it involved the self-organisation of communities. Class emerged as less of an abstraction here than in other eco-activists' presentations. He was also surprisingly sympathetic to Douglass' class analysis of Climate Camp, describing many supporters as weekend hippies but stressed that many did not fit this description. There was a suggestion by an NUM delegate that the union might be prepared to pursue an anti-open-cast collaboration with environmentalists, as it had with 'No Open-Cast' in the 1990s.[6] There was also a lot of general debate between the various presentations, my favourite contribution being a sort of ode to coal as alchemical material by a retired miner, thanking it for the gift of class struggle.

The conference often threatened to become nothing but the conjunction of two forms of reformism – trade unionism and environmentalism – disputing the response to climate change rather than providing a challenge to the commodification of the world that both climate change and capital are predicated upon. Both trade unionists and activists discussed climate change and class as though they were only connected when the poor, or a particular segment of the working class, were victims of disaster or a shift in production, necessitating the intervention of a union or activist community.

Beyond the stereotypes of pit helmets and dreadlocks the central question the conference raised for me is how to formulate a response to climate change capable of resisting capital's own one – given that capitalism loves a good catastrophe from which to extract value. Is there an inherent connection between capital, disaster and labour? In 1951, Italian ultra-leftist Amadeo Bordiga drew on Marx's concept of 'dead labour' (past labour solidified in the infrastructure it has produced) to demonstrate why capitalism is 'the masterful development of an economy based on disasters'.[7] In his words, 'To exploit living labour, capital must destroy dead labour which is still useful. Loving to suck young warm blood it kills corpses'.[8]

His point was that capital thrived on disaster because it provided the opportunity to extract more surplus value from living labour through production. Bordiga suggests in this the way that disaster, capital and labour are imbricated – class and labour rather than being factors to consider in the disaster of climate change are central to it. Climate change often seems to be the product of two inhuman agencies, nature and capitalism, but it's unlikely that a return to trade union forms of organisation, even 'one big union', could produce the necessary oppositional force to counter this. Despite the tensions within the conference, I felt it was constructive in beginning to open a dialogue around this issue: what forms of class composition and organisation might arise within a climate change paradigm dominated by an increasingly authoritarian state and capital's need to prosper?

Info

The Labour Movement Conference 'Class, Climate Change and Clean Coal – the Climate Campers and the Unions' was held at the Bridge Hotel in Newcastle Upon Tyne, 1 November 2008

Footnotes

[1] *Climate Camp Newspaper*, August 2008. Copies can be obtained at, networking@climatecamp.org.uk

[2] Dave Douglass, Climate Camp report, http://www.indymedia.org.uk/en/2008/08/407011.html

[3] For more on carbon storage see, 'Techno-Fixes: A critical guide to climate change technologies', Corporate Watch report, 2008, pp.35-39, http:// www.corporatewatch.org

[4] Will Barnes, 'Capital Climes', *Mute*, vol 2 #5, 2007, http://www.metamute.org/en/Capital-Climes

[5] 'Techno-Fixes', op. cit.

[6] For details on this campaign see, *Do Or Die*, issue 7, pp.23-32, http://www.eco-action.org/dod/no7/23-32.html

[7] Amadeo Bordiga, *Murdering the Dead*, Antagonism Press, 2000, p.31.

[8] Ibid, p.36.

> John Cunningham <coffeescience23@yahoo.co.uk> lives in London and is still looking for a way out

'THE SIMPLE EXPRESSION OF COMPLEX THOUGHT':
FOR A MEDIA THEORY OF EXPRESSION

> In light of the postmodernist cul-de-sac of relativism which, for all its social constructivism, cannot escape crude causality, M. Beatrice Fazi proposes a metaphysics of difference for decoding expression in interactive media

While walking through a recent Rothko exhibition at the Tate Modern, the audio guide I was provided with tells the story of how Mark Rothko disliked the label 'abstract expressionist' assigned to him by the critics because he didn't consider his paintings to be unreal compositions (abstract), nor private accounts of his internal emotional life (expressionist). His big, feathered, flooding patches of colour are, in fact, a subtle dynamic of existing processes and materials. No intimate details of the artist's subjectivity are to be found, but layer after layer of imperceptible paint as 'the simple expression of complex thought.'[1]

This anecdote per se is rather fascinating, for it concisely and interestingly pinpoints two common misconceptions concerning both abstraction and expressivity. Generally speaking, that which is 'abstract' is understood as the non-figurative, the symbolic, the transcendent. Fiercely independent from concrete references to the world, abstraction is, for this reason, denied external reality and confined to an imaginary of internal recursive significations. 'Expression', on the other hand, is the

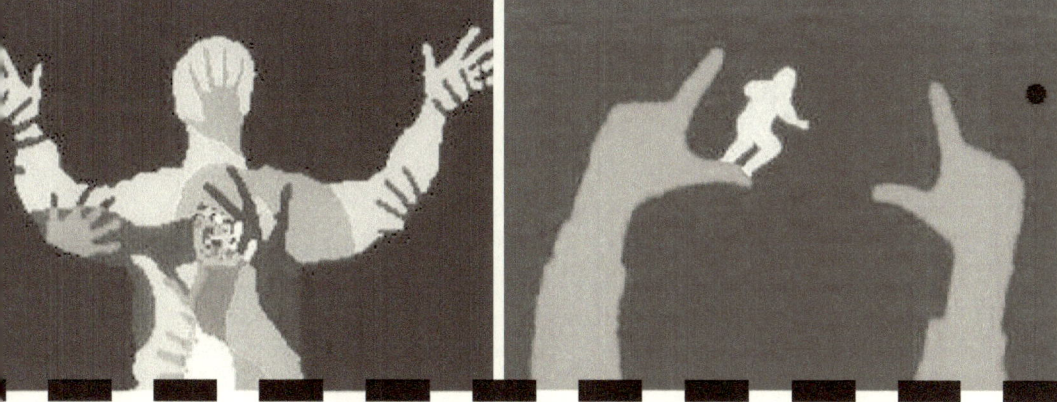

exposed, the spoken, the gestural or, in other words, the communicated qualities of inner experience in all its emphasis and distortion through vivid emotional utterance.

From art history to cultural theory via daily practices of common sense, such denotations seem to stand firm through a large part of contemporary social phenomena, which inevitably involve and develop into discourses on technologisation and digitality. With their capacity for mediation and re-mediation, these latter form the ultimate outposts of communicational practices of mass expressivity on the one hand, or, abstract corporeal decay and immaterial representation on the other.

Translated into practices of 'interaction' (a word so burdened and inflated to have become, as Massumi splendidly comments, a proper regime of tyranny[2]), the 'user' is, from this perspective, a sort of strange creature in between a behaviourist Golem, pushing buttons and waiting for feedback, and a transcendental ego of Kantian reminiscence, positing itself and its relation to technology as a world-creating operation of knowledge. Whether adding a layer in Photoshop or jacking into that fictional dimension formerly known as cyberspace, expressivity and its cognates do not seem to go much further than the limits of their own subjectivising instantiations. In other words, expression becomes a designative rapport between the intentionality of a subject and the phenomenality of such a transmission of information. Interactive practices, in this sense, are organised around the more or less conscious or intelligent will of their actors, whether humans or machines.

According to what can arguably be called a 'communicational' understanding of expression, the user's knowledge and intentions are the basic premises against which to model the semantics of the interactive system.[3] Such levels of cognitive representations, then, are called 'abstractions' for they involve thought – or, better, rationality – as a means to formulate a plan of action and consequently evaluate the results. What is affirmed, in this perspective, is the referential function of expression intended as causal transmission of some sort of content from a subject A to an object B. In turn, 'abstraction' is a tool with which to perform, simplify or support such a communicative task, allowing logical systems to be reasoned out. Whenever, for instance, I decide to upload my holiday pictures on the next best social networking website, my establishing the goal, specifying the intention, executing the action, sequence and finally judging the system's state all become subjective, expressive acts, whose legitimacy and efficacy are guaranteed by a logically antecedent content to be expressed by

All images from Myron Krueger's *Videoplace*, 1974

means of abstract thought processes, where 'ideas' are distanced from their referent 'objects'. This is, indeed, also a quite traditional and very common framework in a large part of Human-Computer Interaction.[4] Designers of software and hardware aim to create computational products for people who have specific tasks in mind and who want to use machines in a way that is supposedly fluid in respect to their functions, so translating 'abstract' knowledge into an executable system. If we now look again at the aforementioned example of the 'upload picture' function of a social networking site, we will see then that it can be said to work exactly in accordance with such HCI principles. Expressivity, thus, is supposedly established by the user's own path through windowing systems, cursor indicators and all the technological features that are meant to present or represent a mirroring or moulding of his/her intentionality.

Admittedly, the parallel drawn here between the homogenised and amateurish constraints of Web 2.0 and the highest spheres of artistic avant gardes of the 20th century could appear bizarre. Yet, the conceptual rationale behind the two is possibly the same: to pursue and interpret expressionism as a twofold movement of interiorisation and exteriorisation, which of course differs from artistic technique to massified technics, but nevertheless presupposes inputs to be translated into stimuli for the system; an interface allowing the interaction to be successful and a hypertrophied centre of subjectivity presiding over the symbolic exchange.

Such a 'fetish' of relations and responses, perpetuating the myth of interaction involving at least two participants (the user and the system), is premised on a self/other dialectical dichotomy that the best poststructuralist contemporary thought has already helped to expose and dismantle. Very few theorists today would disagree that an alternative to the Cartesian-Kantian subject has offered, in recent years, a fortunate intellectual metaphor with which to bridge disciplines and suggest new models of critical analysis. Also techno and software cultures, in this regard, seem to be better understood through more comprehensive environments and composite networks of connections across the machine and the organic. This position rejects the very notion of a human being as a centre of conscious thought and action, and aims to acknowledge the manifold aspects of technological instantiations and their socio-cultural contexts.

In my view, though, it can be said that a postmodern emphasis on the social-constructivist critique of subjective communication – with its self-reflective universes of simulations, reproduction and deconstructed meta-ideologies – has not liberated abstraction from its own image of signification and denotative correspondence, nor the interactive system from its own expectations of usability. Rather, positioning mental images and their relationship to one another in a relativist but yet causal framework results in an anti-metaphysical gesture against the concretion of intelligible structures and formal principles of reality. Such stances – and here is the main point – allegedly compose the whole through an industrious, experimental construction but, in fact, reinstate a correspondence of content between simulation and simulated, produced and reproduced, the user and his/her behaviour.

In this regard, then, one could argue that the postmodern reworking of expression still demands an act of synthesis between the parties. Let us think, for the sake of argument, about media theory questions like how we define ourselves in the digitally networked space and how these have often been explored and expanded as new ways of producing the body of deconstructed subjectivities. New media artefacts, in this sense, are seen as actively producing digital embodiment. Similarly, our relationship to subjectivity and its corporality are supposedly made 'different' through our encounter with digital technology.

To develop this point, let us now take as an example the classical and seminal interactive media art pieces of Myron Krueger and his pioneering mixed-reality environments, and try to interpret them through the aforementioned relativist and postmodern conceptual framework. In Krueger's *Videoplace* (1972-1990s), virtual reality can be experienced without wearing awkward goggles or gloves.[5] The visitor's movements are captured, analysed and outlined in an image of the body as processed by the interaction with the machine and other graphic poles in the system. By adding the body back in the virtual picture and so relating the different to the different, (whether material substratum and computational processing or the condition of action and the conditioned response), the work creates a clever series of responses and linkages by means of composition, articulation and connection.

> 'The world', writes Deleuze, 'does not exist outside its expressions'

In this regard, though, it can be said that such a play of absence and presence recalls a sort of notoriously Hegelian 'labour of the negative' through which interaction sets out to win its freedom. Differences between parts of the interactive system are premised upon a relation of determination in which all the elements – posture, rate of movement, visual and auditory reactions, etc. – must be somehow negated and recomposed by the computer processing in order to attain reality and identity. While it is true that the subject of this system is not intended to be a cognitive unity of perceptions, the expression of action of such a dismantled centre appears to be determined by mechanisms of simulation which invest a lot into the polarity of complete action-response and its final manifestation.

The Simple Expression of Complex Thought

To a large extent, this is also the contradiction emerging between experience and determination in strictly socio-constructivist and relativist models of technological expressionism. Form is dismissed in favour of behaviour with the intention of opposing uncritical and naïve subjectivism and the unity of individual life. From this perspective, practices of interaction seem to evoke patterns and roles of action and conduct them over a predetermined and monolithic mathematical form. Interactive structures – from artistic compositions to commercial products via industrial applications – are said to be grounded in a demanding reality, much richer in behaviour than formal and algorithmic systems, developing the idea that knowledge cannot be passively absorbed but must be actively situated and constructed by the system itself.

In a very 'ubicomp' spirit, today's technological practices appear in fact to bring to the fore the question of how we can interact with conversational situations rather than with single appliances.[6] The goal of a 'calm technology' is thus to engage with the full complexity around and within the system by connecting things in the world via a sort of 'socio-constructional' computation. In this view, for instance, an iPhone is precisely such an amazing piece of computational gadgetry due to its capacity to relate to the elaborate world of its user. Thanks to software components and hardware sensors able to detect and respond to external reality, an object as mundane as a mobile phone then becomes a world-making experiential medium, which creates a situation relative not only to its social functions but also to the user's personal expectations.

expression is a plane of immanence for the production of pre-individual subjectivities

The actualisation of these patterns of interaction are then prevalently preoccupied with overcoming moments of discontinuity, articulating differences in a sort of 'discursive' composition of parts. The design of the computational object – be it a commercial device or a piece of art – begins with an abstraction of what it should do and ends with the behavioural completeness of its contextually constructed or deconstructed conditions. The expression of such a relational system aims to trigger behaviour on both sides of the classical dichotomy object/subject, machine/user. In my view, though, this conceptual outlook has a reductive understanding of both 'form' and 'formality' that could be somehow misleading. First, it seems to assume that information can be arbitrarily opposed to its medium in which it is in fact 'informed'. Second, it looks at technology as a carrier of the narrow utilitarian purpose of creating the 'human' and assesses digital systems as infrastructures for such participative ends. Finally and most importantly for our discourse, by depriving expressivity of an ontological form, it makes of it a function between components whose user-friendly goal is to bridge between collections of interfaces corresponding to desired useful behaviour. Metaphysical apertures are, once again, replaced with an epistemological matrix – even if under a much more fashionable, postmodernist disguise.

But then, how is it possible to explore pathways to expression that would consider the relationship between things as referring not only to the signifying characteristics of the elements in composition or the communicational exchanges between them, but also to the potential of expressivity itself to create these kinds of relation? How, to return to our example, could we make *Videoplace* expressively speak through all the planes of its computational, physical, technological and aesthetic potentiality, without giving in to the lure of moving through a trajectory which is composed or deconstructed in quantitatively relational terms? In this respect we need to ask what happens when the creation of media objects cannot be discerned from their complicity with multiform and immanent life? Have anti-metaphysical stances paralysed heterogeneous agency and complex thought or simply exiled it to a sort of postmodern relativist postmortem? Promises of interaction are common to virtually any rhetoric of technological experience. Similarly, expressivity is a sort of Holy Grail for any artistic practice worthy of the name. The aesthetic role of expression, I believe, can in this sense be understood and analysed in terms of modes of relation to multiplicity and creative production. To what extent the mutual implication of practice and theory can inform such research is not just a question of methodological framework but, principally, an ontological view.

More broadly put, theories of interaction can be philosophically investigated as operating through a logic of differentiation. The role of expression is thus to think what remains unthought of 'difference' – a concept so profoundly charged with logical, ethical, aesthetic and ontological implications – under the conditions of symbolic models. This expressive value of differentiation is 'outside' the strict domain of representation but not necessarily in opposition to it. Real difference and real change, then, preside over the division between extension and intensity and require us to

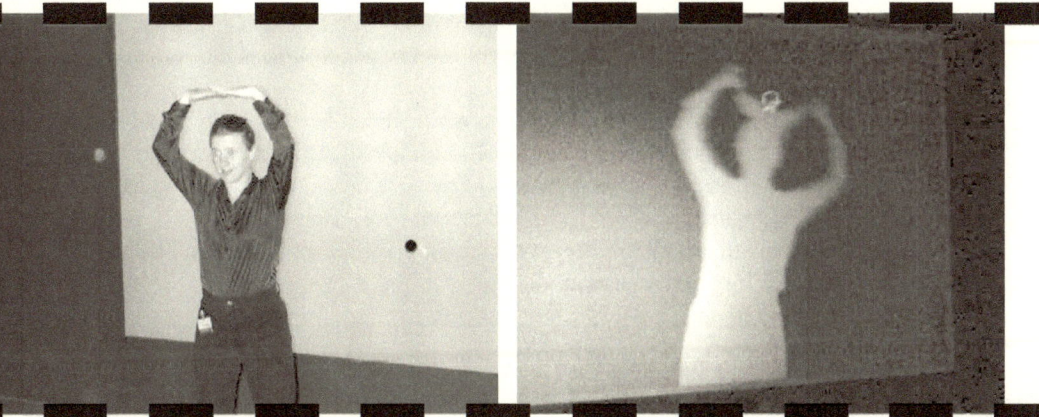

Image: A person interacting with *Videoplace* at the Prix Ars Electronica exhibition, Austria, 1999

The Simple Expression of Complex Thought

look for a differential ground of agency that, in turn, never casually resembles the grounded practices. The unthought of difference, in this respect, becomes the metaphysical condition of what there is. In other words, this dimension of differentiation is an impersonal and pre-individual plane that does not correspond to an empirical field as such but that makes it possible. Consequently, such a characterisation of experience and the experienceable leads also to a reconsideration of our understanding of abstraction and images of thought. The possibility of thought is not to be confused with models of recognition, but is understood as the product of processes or, more radically, as a process itself. The 'expressive problem', in this respect, can be developed as an 'anti-redutionistic' project, opening up fields for new and unexpected aesthetic practices.

Arguably, a truly innovative philosophical system is defined by its repositing of problems through a new array of concepts, which inform and forge the thought itself. Over relatively recent years, the late philosophical figure of Gilles Deleuze has established himself within Anglo-American academia as a leading voice in critical theory in virtue of the resource of original concepts his philosophy seems able to provide, not only to scholars but also practitioners. Discourses on the 'rhizomatic' nature of our culture, 'machinic assemblages' and 'affective intensities' have flourished from biology to visual art, most often successfully engaged and engaging yet sometimes repeated ad nauseam. While I have no intention of reciting here this structure of thought – if we can ever talk of 'structure', in fact – it is nevertheless central to my argumentation to briefly linger on his study of difference as a logic of the ontological distinction of degrees. Such a strictly metaphysical part of the Deleuzian conceptual 'toolkit' has, of course, been acknowledged by academia and contemporary philosophy. Nevertheless, in my view, its dynamic formality and its ontological legacy – which are somehow still quite unexplored in respect to more immediate and conceptually less demanding parts of Deleuze's thought – offer an immense potential also for speculative practices, especially in their technological instantiations.

> **The activity of differentiation is an expressive agency, a movement, a force**

Within the academic debate, the philosophical tradition I am referring to has been known as the 'philosophy of expression'. Its strength, though, is in its not having been formalised through a specific school of thought, but in its remaining transversal to different times, logics, aesthetics and metaphysics – from Spinoza, to Nietzsche via the Stoics, Leibniz, Bergson, some Scholastic philosophers and finally arriving at the Deleuzian formulation.

In this respect, 'expressionism' is an ontological attitude, rather than a system, a philosophy of emergence, affirmation, creation, forces, change. 'The world', writes Deleuze, 'does not exist outside its expressions'.[7] In his radical and original technique of reading philosophical texts against each other, the aforementioned philosophical strands and personalities vibrate and resonate, one by one, and against each other. Questions about abstract materiality, forces in composition and complex agency are interwoven into an ontological and aesthetic discourse of the expressivity of the real 'otherness'. This sort of Foucauldian 'space of the outside' has had many names in the history of philosophy, yet in this context we will refer to it simply as the search for the 'other' perspective, which does not put the human being or its conscious activity at the centre of the construction of reality.

In particular, Deleuze makes of expression the very core of the Spinozist ontology. The idea of expression is nowhere explicit in Spinoza's work, yet it operates silently, between the lines, as the motor of ontological differentiation. Deleuze calls 'expressionism in philosophy' a triadic problem:

> we must distinguish what expresses itself, the expression itself and what is expressed. [...] 'What is the expressed' has *no existence* outside its expression, yet bears no resemblance to it, but relates *essentially* to what expresses itself as distinct from expression itself.[8]

Substance – famously depicted by Spinoza as one, infinite and univocal – expresses itself as *causa sui* through the attributes. Similarly, the essence of the Substance is characterised as that which is expressed. In other terms, expression is a plane of immanence, the intensive condition for the production of pre-individual subjectivities. Reality is viewed as an expressive means of understanding itself as intensive power that expresses itself through activity. Yet, having no semantic, communicational or even contextual resemblance to expression, the expressed is, *qua* essence of what expresses itself, distinct from expression itself.

Deleuze both maintains and corrects the Spinozist framework and uses it to develop his own realist and anti-reductionistic ontology. All philosophy of expression is, in this sense, a philosophy of difference. The activity of differentiation is an expressive agency, a movement, a force – a reality of implications and explications according to which experience is a pre-linguistic and pre-representational activity that can not be separated from the act by which it is expressed. Contra phenomenological understandings of expressive languages as intentionally directing a psychological representation via an abstract content to an ideal or physical referent, Deleuze's expression is neither communication nor socially relativist symbolic exchange. Similarly, it is not a 'relation' per se: within the horizon of the univocity of Being, the consistency of relational structures is independent from the terms which affect them.

The Simple Expression of Complex Thought

Expression relates as much to actual things, which have an identity, as to virtual ones, pure variations. Such a philosophical struggle against dualism and dialectical reductions considers abstraction a form of reality as much as experience – neither physical or corporeal but yet real and existing, in evolution with other materialities. It folds and unfolds in response to forces around it. Immanence is, in this sense, as much a principle of thought as of action. This does not mean that we should emphasise sensible experiences – together with introspection – as the only possible knowledge. Interpreting reality as an asymmetric, dynamic relation of both abstract and actual intensities means looking at the world according to its virtual tendencies, contra the transcendental illusion that does not respect such processes. Reality thus lies in the disparities between its orders, in the 'differential'. This virtual composition specifies an account of determination that eludes systems of essentialist classification and does not relegate difference to a negative position in respect to identity.

But how can this philosophy of expression speak to media theory? How can expressionism be used as a conceptual tool to think technology beyond questions of knowledge and representation, and productively engage with symbolic systems, whether as a culturally generated set of events or as a natural experiential field of both agency and thought?

The realist ontological approach to media studies that I am proposing here does not intend to stage a revolutionary act or to imply some sort of Heideggerian concealment operating in the computational world. Rather, it uses different concepts, practices and philosophical positions as 'windows' into speculative realms of thought and action, clarifying machines of interaction as machines of expression – which are not necessarily 'subjective' or 'constructed', like the paradigms mentioned above. Translated into a media theory context, the representational power of the computer – procedural, encyclopaedic, participatory and spatial – is forced to face the question of how it can be unfolded into further layers, vectors, glitches, dynamic forms, tangents and nodal points.

Talking of metaphysics in digital culture is not, by many accounts, a popular move. As a matter of fact, any metaphysical approach is usually regarded as a sort of 'positivist insult', standing for abstruse speculation and irrational clutter.[9] In particular, in computer science 'ontology' is understood in its most Aristotelian determination, namely the study of how entities can be categorised and associated in hierarchies according to similarities and differences. In this regard, the language of logic conceptualises the domain through methods of classification, unfolding the terminological and conceptual incompatibilities, which inevitably arise within such a formalisation. Consequently, the relationship between the abstract, logical concepts underlying computer operability and their physical realisation is said to be a representational one and computer science per se can be considered as an attempt to empirically explain what symbols are.[10]

According to the aforementioned Spinozist-Deleuzian articulation of expressionism put forward here, though, levels of logical and physical abstraction are not meant to epistemologically distance coding processes from their mundane manifestation of bits, voltages and wires. Rather, such an expressionist position looks at the emergent properties of computation – belonging to the interaction between parts – as accumulations of both hardware and software operations. In this regard, the conceptual and practical possibilities that a theory of expression opens up in relation not only to new media studies but also to the poetics of interactive media and their socio-cultural context are manifold. The adventure of determination is not just a method with which to critique representation. Such a genesis of form does not reduce natural complexity into cultural simplicity but, instead, relates to the immanent capacities of media to give rise to flows of matter-energy in-formation. Interactivity, in this sense, is always a case of encounters: real layers of abstraction, thought, agency, materials and principles that compose the interaction as a differentiated and differential immanent field of being. Informational objects unfold as objects of a problematic, dynamic and differential form – an intrinsic mode inevitably drawn to 'interaction' as an affirmation of life in all its complexity.

Media objects have not been waiting around for their 'users' in order to exist

This realist metaphysics attempts to cut across phenomenological, subjectivist, humanist, transcendent and relativist understandings of interactivity and especially substitute questions of 'signification' with problems of 'significance'. Media objects have not been waiting around for their 'users' in order to exist. Users do not assign signification to things, for these are already in a network of object relations, independent and before cognitive subjects even appear. If we want to use a metaphor, a piece of software code expresses itself like a DNA string would do: it is abstract yet real, having concrete effects and emerging physicality. This does not mean that the world needs to be coded in signifiers to have form. Quite the opposite, in fact: form – technological, biological, linguistic, cultural, etc. – does not come from an 'outside', is not imposed on an inert receptacle in a cosmological and transcendental fashion. The materiality of computation is already 'morphogenetically' pregnant, capable of generating and expressing form.

Just to provide a small example, let us think about digital image compression, whose aim is to cut on data redundancy so as to transmit and store information efficiently. This compression can be 'lossy' or 'lossless'. In particular, lossy compression methods introduce compression artefacts in order to produce a considerable reduction

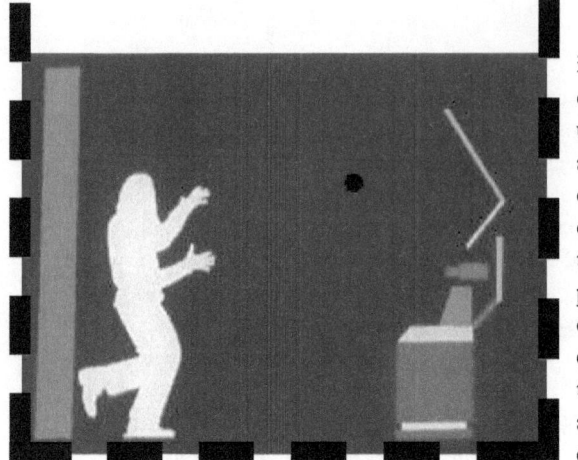

in bit rate. From a technical point of view, such artefacts are a particular class of data error that is usually the consequence of quantisation. Of course, repeatedly compressing and decompressing the image will cause the file to progressively lose quality. Yet we could note that the machine expression takes place exactly in the loss. In other words, how the algorithm selects what to lose is exactly the site of its expressivity. This 'decision', obviously, is not organised around an epistemic centre of the machine's will. Expression, indeed, is always *of* a relation: content and meaning (namely, the coding of the compression) are inseparable from agency and execution, yet they do not resemble them. The computational continuity of such expressivity is, once again, a generative problem of interaction, assured through the actualisation of the pragmatic dimension of language.

In this regard, the great challenge for a media theory of expressionism is precisely to engage with representational systems. What is really at stake, in this sense, is the role of computation, coding that exceeds its discreteness and becomes 'difference'. Again, the question for media theory is to think how the non-representational dimension of technological experience can undertake the symbolic of the machine. How, in this respect, we can account – with words, logics, codes or anything else – for a heterogenic expressivity that can travel through all the scales and degrees of such an infinitely transformable and mutable structure we call nature. Of course, technology does not happen in a vacuum. Yet, in order to avoid the phenomenological risk of reinstating a 'dative' in the generation of the sensible, it is also necessary to posit a nature-culture continuum. Distinctions between what is perceived as 'natural' and what is instead 'cultural' are no longer possible, if they ever were. Media are cultural objects that appear as natural problems. They are invented, fabricated, created, yet their happening is not limited to the reality of what is 'constructed' but extends to what we can intrinsically relate to as the 'given'. Technology, in other words, is always a vehicle for exploring and expanding culture. Such a cultural dimension, though, lies outside of a particular relativist context and emerges as a natural stratification of practices, thoughts, processes, bodies, mutations and functions. A car being put together on the assembly line, for instance, becomes from this perspective the natural yet techno-cultural expression of the history of its materials, the processuality of human inventions, the routine of working shifts and effectiveness of political decisions, just to name but a few of the conditions of possibility that can not be understood as something innate or predestined for the representational cogito.

To conclude, an expressive metaphysics of media ecologies is not only a battle against the neurosis of teleological subjectivism and the paranoia of technological relativism. It is a proper media ontology having both an intensive 'inside', which is more broad and profound than cognitive interiority, and an extensive 'outside', which does not stop at the exteriority of sensible perception. Here is, to close a circle we opened with art history and developed through communication theory, postmodern critique and continental philosophy, 'the simple expression of complex thought', or a media world that really does not exist outside its expressions.

Footnotes

1 From Rothko's joint manifesto with Adolph Gottlieb, originally published in *The New York Times* on 13 June 1943. See James E. B. Breslin, *Mark Rothko: A Biography*, Chicago: University of Chicago Press, 1993, p.193.

2 Brian Massumi, 'The Thinking-Feeling of What Happens: A Semblance of a Conversation', in Joke Brouwer and Arjen Mulder (eds.), *Interact or Die!*, Rotterdam: V2_Publishing/NAi Publishers, 2007, pp.70-91.

3 See also Brian Massumi, 'Introduction: Like a Thought', in Brian Massumi (ed.), *A Shock to Thought: Expression after Deleuze and Guattari*, New York: Routledge, 2002, p.xv.

4 Broadly speaking, HCI is defined as the field of computer science that studies the direct link between people and computers.

5 See Myron Krueger, 'Artificial Reality: Past and Future', in Sandra K. Helsel and Judith Paris Roth (eds.), *Virtual Reality: Theory, Practice, and Promise*, Westport, CT: Meckler, 1991, pp.19-25.

6 In the late 1980s, computer scientist Mark Weiser envisaged a future of small and distributed computational devices seamlessly integrated with the environment in all the scales of everyday life. See Mark Weiser, 'Some Computer Science Problems in Ubiquitous Computing', *Communications of the ACM*, vol.36, no.7, pp.74-83.

7 Gilles Deleuze, *The Fold: Leibniz and the Baroque*, trans. Tom Conley, Minneapolis, MN: University of Minnesota Press, 1993, p.132.

8 Gilles Deleuze, *Expressionism in Philosophy: Spinoza*, trans. Martin Joughin. New York: Zone Books, 1992, p.333.

9 Famously, the early 20th century Vienna Circle is said to have rephrased Wittgenstein's stark statement 'whereof one cannot speak, thereof one must be silent' into the imperative 'Metaphysicians: shut your traps!'.

10 This is a classic position in AI, stated by Newell and Simon in their work on physical symbol systems. See Allen Newell and Herbert A. Simon, 'Computer Science as Empirical Enquiry: Symbols and Search', in Margaret A. Boden (ed.), *The Philosophy of Artificial Intelligence*, Oxford: Oxford University Press, 1990, pp.105-132.

M. Beatrice Fazi <b.fazi@me.com> studies humans and machines and finds both of them quite beautiful. She is currently a PhD candidate, in the field of interactive aesthetics and continental philosophy, at the Centre for Cultural Studies, Goldsmiths

The Passive Vampire, Romanian surrealist poet Ghérasim Luca's recently translated book, brings objects and desires into intimate contact, with unexpected results. Review by Kenneth Cox

OBJECTIVE PHANTOMS

Ghérasim Luca's *Le Vampire Passif* has for many years been surrounded by an aura of mystery, like a 'forgotten' or lost *grimoire* of surrealist writing; a book that, like its author, has had something of a 'phantom existence'.

First published in Budapest in 1945, by the appropriately named Éditions de l'Oubli ('Forgotten Books') – written in French, and not his native Romanian – in an edition of only 460 copies, the book was not republished until 2001, by José Corti. It was not only its inaccessibility that created the book's legendary status, and not only amongst that minority readership interested in surrealism, but the personality of its author.

Born Salman Locker in Bucharest in 1913, the son of a Jewish tailor, it wasn't until much later that he became Ghérasim Luca. Drawn to poetry and the Romanian avant garde, when Locker was about to publish his first text, the pseudonym 'Ghérasim Luca' was suggested by a friend, which he promptly used. Only later did he learn that his friend had stumbled across this name by chance in an obituary. In 1940, together with Gellu Naum, Dolfi Trost, Virgil Teodorescu and Paul Paun, he founded the Romanian Surrealist Group, following a visit to Paris in 1938 where, through fellow Romanians Victor Brauner and Jacques Hérold, he had met the French surrealists – a decisive encounter. Although in existence for only seven years, cast adrift in clandestinity throughout WWII and mostly remaining in obscurity ever since, the Romanian Surrealist Group was nonetheless one of the most explosive, original manifestations of surrealist thinking and practice. Throwing down the gauntlet with their pivotal statement, composed by Luca and Trost, *The Dialectic of Dialectic: a Message to the International Surrealist Movement*, they set out to challenge any slide into complacency that might result

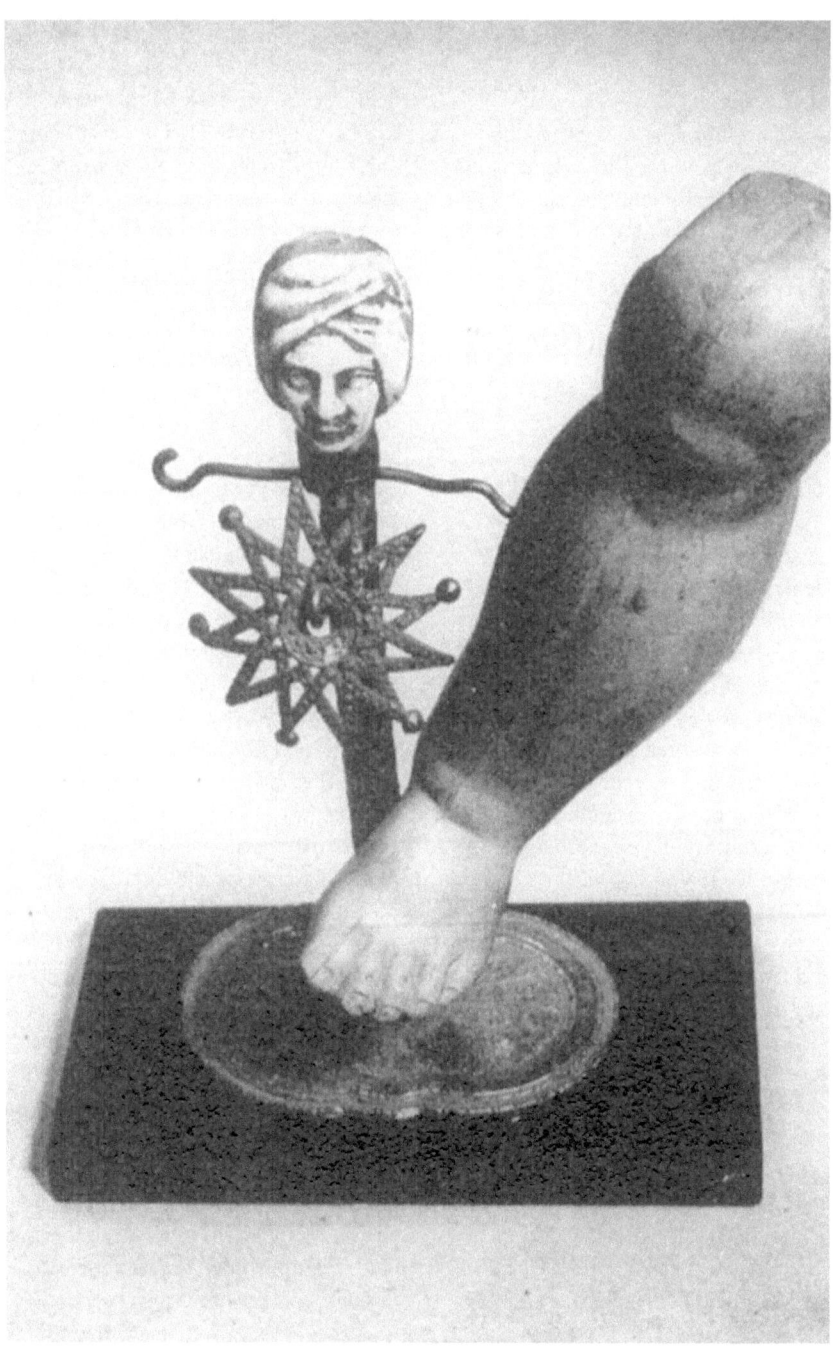

Image: Ghérasim Luca, *Déline-Fetish*

Objective Phantoms

in surrealism's discoveries being absorbed into means of cultural production only, to prevent it 'sinking into a hackneyed Romantic idealism'. Their declaration makes the uncompromising

> **Luca expresses a desire to transform the world, in an experience of *convulsive beauty***

demand that surrealism remain in a condition of perpetual revolution, through taking a radically dialectic standpoint of continuous negation, and further, the negation of that negation.

In 1952 Luca moved from Bucharest to Paris, the city he loved, where his poetic researches were mostly of a solitary nature. Perhaps best known for his poetry, with its hermetic wordplay exploring the morphology of language, breaking down and rearranging its constituent parts to uncover new meanings, Luca was described by Gilles Deleuze as the greatest living poet writing in French. Having spent over 40 years in France without papers, he was evicted from his apartment in 1994, along with all the building's tenants, victims of 'urban renewal'. At the age of 80 and unable to accept his new situation, Luca committed suicide on 9 February 1994 by drowning himself in the Seine – a poet's death, his body being found exactly one month later, an event curiously foreshadowed in a text from 1945, *La Mort Morte* (Dead Death), composed of five 'suicide notes'.

The Passive Vampire itself is an *object* that is incredibly difficult to describe, as elusive as its subject matter, being a concoction of theoretical enquiry and deeply personal observation, mixing poetic prose and psychoanalytic investigation. On the surface, it deals with the creation and exchange of (highly personal) surrealist objects, illustrated throughout with enigmatic photographs, presented as pictorial evidence in such a way as to place the book in a lineage stemming from André Breton's *Nadja*. In places it possesses a distinct lyrical quality, most likely inspired by Lautréamont, but rather than taking a delirious plunge into the imagination's depths through any purple prose, Luca writes with a disarming honesty and directness in describing and interpreting events.

The book falls into two distinct sections, the first of which is concerned with what Luca terms the 'Objectively Offered Object' (OOO), and describes the circumstances surrounding a number of these composite surrealist objects, each made by combining found or chosen individual items. These composite objects were made by Luca in order to be given, as a means of revealing the hidden relationships between subjects, through an 'active collective consciousness' that is very much analogous to dream. The giving of such OOOs is differentiated from the giving of ordinary presents, an act which has been reduced to mere convention or habit and from which the force of desire has been drained. These objects, on the other hand, are made as vessels for desire and as a means of deciphering unconscious messages, which for Luca are signs that, in their combinations and interpretations, primarily carry highly charged erotic meanings. The OOO is thus somewhat like a magical spell that both describes a desire and at the same time reveals it, and even perhaps *invokes* it, as if causing an event to happen. And as the truism goes, you should be careful what you wish for.

One such object, entitled 'The Letter L', is constructed from an old, wooden child's doll found in an antique shop, with hundreds of pictorial riddles from the pages of an almanac randomly pasted over its torso and leg, and with another doll's head disturbingly attached upside down on its groin. Razor blades are inserted into this second doll's head, with one sliced into an eye. The photographs immediately call to mind the violent re-articulations of Hans Bellmer and, more recently, the Chapman brothers. Through associations with *Nadja*, this object had been made as an embodiment of Luca's desire to form a rapport with André Breton, whom he admired and had met only once, briefly. As Luca expresses it:

> The doll found in the shop window and the envelope full of riddles in the drawer only imposed their presence, violently, into my life at the moment when the desire to know B. [Breton] located in them the overt substitute means for doing this. The incubus found its full realisation through the use of these two magic objects in which I was also shortly to discern sorcery's demonic power. (pp.44-45)

There is something distinctly sulphurous in Luca's allusions, from his poetic hermeticism to the various thaumaturgical and satanic references that run through the book. Certainly, there was a ritual element to the creation of these objects, doubtlessly stemming from his participation in various collective games of the Romanian Surrealist Group; games of giving and receiving 'awards' in absurdist ceremonial, and those of exploring the poetic qualities of objects in a darkened room through touch alone. These were games without competition, based upon exchange and complicity, without a predetermined point of arrival; through play the participants were able to explore the relationships that exist between subject and object, and the latent messages that are carried by the objects through a web of inter-subjectivity, in a 'language of desire'.

A striking passage gives an account of an earthquake one night in the streets of Bucharest, a description that is both objective and oneiric, mediated by another

Objective Phantoms

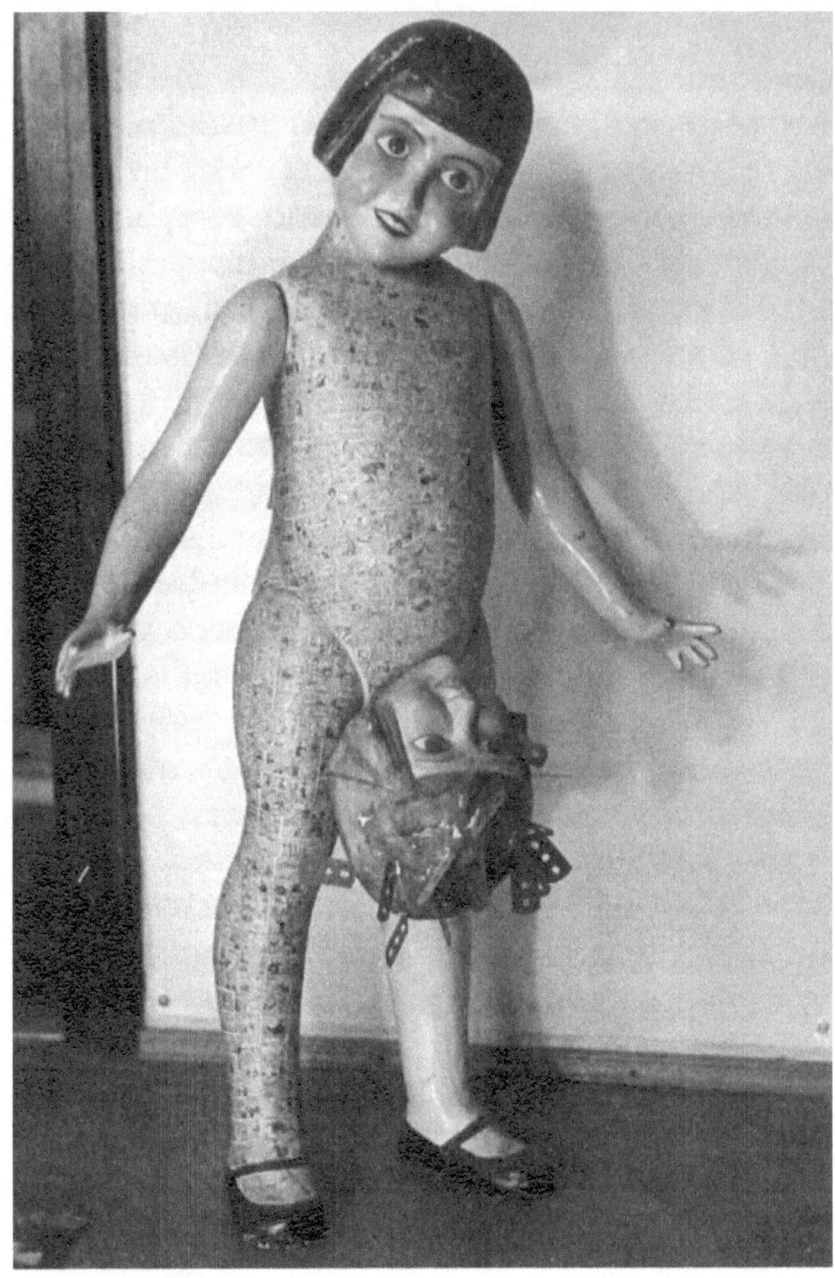

Image: Ghérasim Luca, *The Letter I*

Kenneth Cox

OOO, entitled 'The Ideal Phantom', which in Luca's interpretation brought together two subjects in the most extraordinary of experiences:

> I was awoken at 4 a.m. by a dreadful earthquake: the walls were shaking, wardrobes flung across the room, books falling down on all sides, objects and glasses smashing. Throughout the duration of the quake I kept shouting that I knew it would happen. These powers of prediction, which I was discovering for the first time, only increased my terror. Half an hour later G. came over from the other side of town to see if I had survived, and told me the city was in ruins. I gave him The Ideal Phantom, and we went outside. The streets were full of destruction and rubble, and this town I'd never liked, with its stupid people, stupid streets, and stupid houses, was now unrecognisable, now it had a truly unique beauty, and scantily clad women traversed it like ghosts. (pp.56-59)

Luca expresses a desire to transform the world, even if this transformation was to be brought about by catastrophe, in an experience of *convulsive beauty*. A recombination of objects and subjects for Luca might even act as a precipitate of desire, provoking a genuinely revolutionary convulsion of reality that is inextricably bound up with social and political transformation and, one might say, without which revolutions that are simply concerned with a rearrangement of power relationships are doomed to failure. This very much holds true in our own turbulent times.

the object becomes a dark lantern illuminating internal and external realities

Following the poetic-scientific accounts of the OOOs, the second section of the book is entitled 'The Passive Vampire'. This second section is more lyrical, its almost satanic litanies going beyond the psychoanalytic into more esoteric meditations upon the relationships between the self and the object, and eventually between existence and non-existence. Luca even contemplates objects as communicating *between themselves*, with the subject – like a phantom or *passive vampire* – attempting thereby to discern those mysterious, invisible connections that are

Objective Phantoms

present in the universe, but are revealed in rare moments when the conditions are right. Such objects have a magical power to effect change, whether consciously directed or as the instrument of unconscious forces. From the process of bringing together component objects into an *assemblage*, Luca is able to solve the magical cipher of desire, at which moment the object is then transformed into a vehicle for the desire invoked, capable of bringing that desire to reality through the revolutionary force of love. The object thus becomes a dark lantern illuminating internal and external realities, bringing together unconscious and conscious in a surreality. Luca's thinking might be dismissed as wild and wishful, but this is poetic thought at work in its most unrestrained form, striving to grasp the workings of objective chance, striving to discover a new language even:

> a new language that genuinely expresses the psychic phenomena which resemble, but are not identical to, dream. This dream which, even if still opposed to external reality, has long since ceased to be opposed to the life of the dreamer. In this language, the one I have been unable to find, the ancient antinomies, beginning with that of good and evil, will be resolved for the meanwhile at an individual level. (pp.78-81)

The book closes with one final, extended account of an exchange of objects, in which an object made by Luca appears to deviate from its intended function as events unfold, to take control even, in a manner that has ill-fated consequences for its creator. The object, entitled 'Déline-Fetish', constructed from a doll's leg, a 12-pointed star and a turbaned head on a metal stand, came to embody Luca's desire for a woman, Déline, with whom he had fallen deliriously in love. But rather than leading to the realisation of Luca's desire, the object somehow seems to bring about a baleful rupture between the couple. It is a poignant account of love that flares up violently and is abruptly lost, leaving us in darkness, on the cusp of Luca's despair.

There is no doubt that *The Passive Vampire* should take its place amongst the essential 'classics' of surrealism's history, but, moreover, it provides a valuable stimulus for any current investigations into the workings of chance and its objects, of dream and desire. As the translator, Krzysztof Fijalkowski writes in his excellent introduction, this work is important 'as a fixed marker for the questions asked today by those wishing to situate themselves in the continuing stream of a critical surrealist thought.'

Info

Ghérasim Luca, *The Passive Vampire*, translated and with an introduction by Krzysztof Fijalkowski. Prague: Twisted Spoon Press, 2008

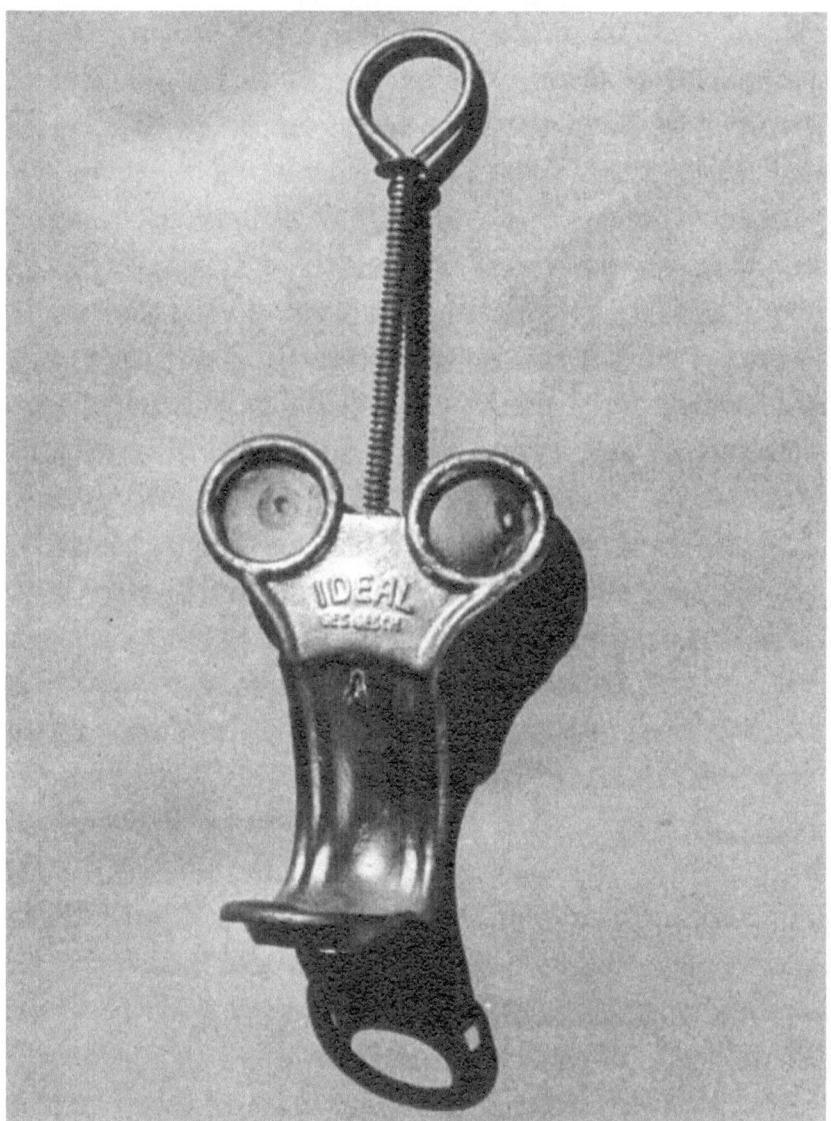

Image: Ghérasim Luca, *The Ideal Phantom*

Kenneth Cox <surrealism@madasafish.com>, a founder member of Leeds Surrealist Group, is editor of the magazine *Phosphor* and co-directs the Surrealist Editions imprint: http://leedssurrealistgroup.wordpress.com

The Mute Archive - a special box set of Mute magazine back issues

Original copies of nearly every issue* of Mute from 1994 to the present - 41 issues, including special inserts, CDs, software and artworks

Mute has been publishing on culture, politics, and technology since 1994, earning an international reputation for originality, humour and intelligence. The first incarnation of Mute was a barely disguised replica of the Financial Times, printed on the same distinctive pink newsprint. Since then Mute evolved into a colour, highly graphic magazine, and finally into the hybrid print/web publication it is today. Featuring many of the artists, writers and photographers who came to epitomise London's status as a creative hotbed, the Mute Archive is a great addition to an academic or personal library.

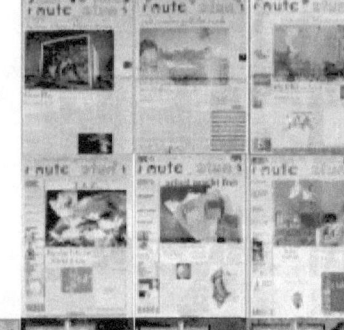
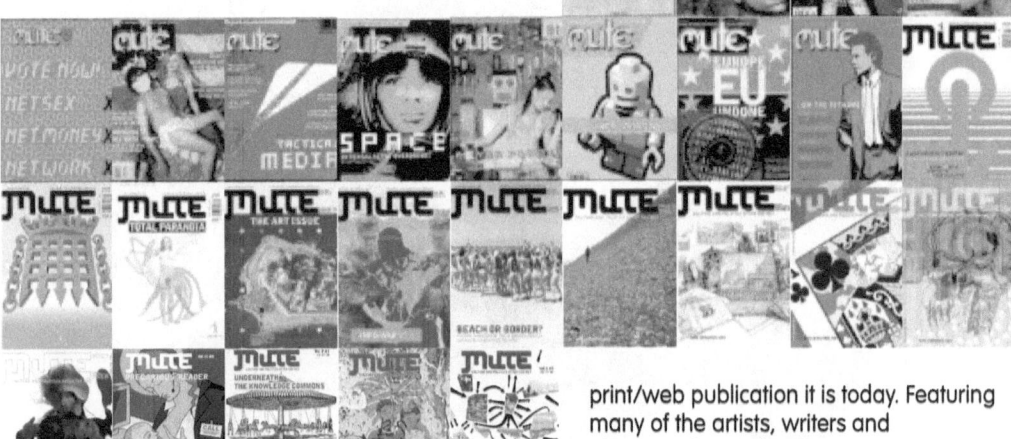

The Mute Archive includes:

The Broadsheet: Issues pilot - 7 (1994-1997)
The Glossies: Issues 8 – 24*
Coffee Table: Issues 25 – 29
POD: Volume II, issues 0 - 11

£200 + p&p
Order online at metamute.org/mutearchive

Institutional and credit card phone orders contact +44 (0)20 7377 6949 Skype mute.london or email lois@metamute.org for enquiries

*we regret that issue 9 is sold out

Magazine Subscription

Mute

Get Mute delivered to your door for one year and guarantee to be first in line for our quarterly collection of provocative articles on culture, politics and technology

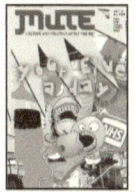 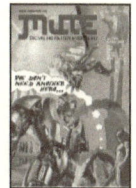

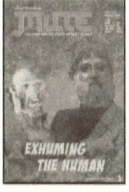 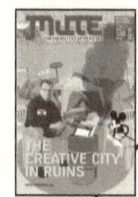

Subscription rates	individual		institutional/company	
	4 issues (1 year)	8 issues (2 years)	4 issues (1 year)	8 issues (2 years)
uk	☐ £20	☐ £38	☐ £35	☐ £67
europe	☐ £22	☐ £41	☐ £38	☐ £72
usa/ rest of world	☐ £25	☐ £46	☐ £43	☐ £82

Please tick the appropriate box.

I wish to pay by cheque/credit card.
☐ I enclose a cheque (GBP) made payable to Mute.
☐ Please charge my
☐ Visa ☐ Access ☐ Mastercard ☐ Switch
Card no. ☐☐☐☐ ☐☐☐☐ ☐☐☐☐ ☐☐☐☐
Expiry date ☐☐ / ☐☐
[Switch only] Issue number ☐☐   Start date ☐☐ / ☐☐
Security code ☐☐☐
Signature _____

name _____
address _____
_____
_____
town/city _____
post code _____
country _____
tel _____
email _____

INSTITUTIONAL OPTIONS:
W: metamute.org/subs
T: +44 (0)20 7377 6949
E: lois@metamute.org
A complete archive of Mute back issues is available

ADDRESS CHANGE:
If you are an existing subscriber needing to change your address, then please email us at lois@metamute.org

Subscribe online at metamute.org/subs or call our credit card hotline on +44 (0)20 7377 6949

Gift Subscriptions
If you are giving Mute to a friend, you can leave their details on completion of your purchase together with your own payment details.

Order online   metamute.org/subs